The Power of Place in Place Attachment

This book provides geographical perspectives on the complex and multifaceted relationship between people and their lived environments. Scholars with varied regional, theoretical, and topical specialties offer chapters that explore different aspects of a phenomenon so pervasive that no conception of social or political action can afford to ignore it. In the process of spatial organization and differentiation, people develop emotional attachments to specific places, as well as people, objects, and practices associated with those places. Place attachments thereby shape everyday routines (e.g., routes to work, shopping, and social interactions), major life choices (e.g., places of residence, education, and vacations), and identities (e.g., civic, national, and religious). These attachments occur across multiple scales from personal dwellings to community, region, and homeland. It is our hope that this book reveals synergies between geography and other disciplines engaging with place attachment whilst invigorating research on the topic. *The Power of Place in Place Attachment* will be of great value to researchers and scholars of geography, identity, mobility, and urban landscape change. The chapters in this book were originally published as a special issue of *Geographical Review*.

Alexander C. Diener is Professor of Geography at the University of Kansas, Lawrence, USA. He works at the nexus of political, social, economic, and cultural geography, engaging topics such as geopolitics and borders, identity and migration, development and mobility, and urban landscape change.

Joshua Hagen is Dean of the College of Letters and Science at the University of Wisconsin-Stevens Point, USA. His areas of research include borders and border theory; cultural politics of place names; demography, economics, and sustainability; and geographies of the pre-modern world.

The Power of Place in Place Attachment

Edited by
Alexander C. Diener and Joshua Hagen

Routledge
Taylor & Francis Group
LONDON AND NEW YORK

First published 2023
by Routledge
4 Park Square, Milton Park, Abingdon, Oxon, OX14 4RN

and by Routledge
605 Third Avenue, New York, NY 10158

Routledge is an imprint of the Taylor & Francis Group, an informa business

British Library Cataloguing-in-Publication Data
A catalogue record for this book is available from the British Library

ISBN13: 978-1-032-43449-0 (hbk)
ISBN13: 978-1-032-43450-6 (pbk)
ISBN13: 978-1-003-36738-3 (ebk)

DOI: 10.4324/9781003367383

Typeset in Minion Pro
by codeMantra

Publisher's Note
The publisher accepts responsibility for any inconsistencies that may have arisen during the conversion of this book from journal articles to book chapters, namely the inclusion of journal terminology.

Contents

Citation Information vi
Notes on Contributors viii

1 The power of place in place attachment 1
Alexander C. Diener and Joshua Hagen

2 Cognitive mapping as a method to assess peoples' attachment to place 6
Jeffrey S. Smith and Ricardo Aranha

3 Rurality as a key factor for place attachment in the great plains 27
Andrew Husa and Cheryl E. Morse

4 Whose Puget Sound?: Examining place attachment, residency, and stewardship in the Puget Sound region 46
David J. Trimbach, Whitney Fleming and Kelly Biedenweg

5 "Tied to the land": Pipelines, plains and place attachment 66
Christina E. Dando

6 Indigeneity, displacement, and regional place attachment among IDPs from Crimea 86
Austin Charron

7 The role of *nostalgia* in (re)creating place attachments for a diasporic community 103
Holly R. Barcus and Amangul Shugatai

8 Multi-scalar territorialization in Kazakhstan's northern borderland 125
Alexander C. Diener

9 Places of memory, historic preservation, and place attachment in Nazi Germany 147
Joshua Hagen

10 Geographies of place attachment: A place-based model of materiality, performance, and narration 171
Alexander C. Diener and Joshua Hagen

Index 187

Citation Information

The chapters in this book were originally published in the journal *Geographical Review*, volume 112, issue 1 (2022). When citing this material, please use the original page numbering for each article, as follows:

Chapter 1
The power of place in place attachment
Alexander C. Diener and Joshua Hagen
Geographical Review, volume 112, issue 1 (2022) pp. 1–5

Chapter 2
Cognitive mapping as a method to assess peoples' attachment to place
Jeffrey S. Smith and Ricardo Aranha
Geographical Review, volume 112, issue 1 (2022) pp. 6–26

Chapter 3
Rurality as a key factor for place attachment in the great plains
Andrew Husa and Cheryl E. Morse
Geographical Review, volume 112, issue 1 (2022) pp. 27–45

Chapter 4
Whose Puget Sound?: Examining place attachment, residency, and stewardship in the Puget Sound region
David J. Trimbach, Whitney Fleming and Kelly Biedenweg
Geographical Review, volume 112, issue 1 (2022) pp. 46–65

Chapter 5
"Tied to the land": Pipelines, plains and place attachment
Christina E. Dando
Geographical Review, volume 112, issue 1 (2022) pp. 66–85

Chapter 6
Indigeneity, displacement, and regional place attachment among IDPs from Crimea
Austin Charron
Geographical Review, volume 112, issue 1 (2022) pp. 86–102

Chapter 7

The role of nostalgia *in (re)creating place attachments for a diasporic community*
Holly R. Barcus and Amangul Shugatai
Geographical Review, volume 112, issue 1 (2022) pp. 103–124

Chapter 8

Multi-scalar territorialization in Kazakhstan's northern borderland
Alexander C. Diener
Geographical Review, volume 112, issue 1 (2022) pp. 125–146

Chapter 9

Places of memory, historic preservation, and place attachment in Nazi Germany
Joshua Hagen
Geographical Review, volume 112, issue 1 (2022) pp. 147–170

Chapter 10

Geographies of place attachment: A place-based model of materiality, performance, and narration
Alexander C. Diener and Joshua Hagen
Geographical Review, volume 112, issue 1 (2022) pp. 171–186

Notes on Contributors

Ricardo Aranha, Department of Geography and Geospatial Sciences, Kansas State University, Manhattan, USA.

Holly R. Barcus, Department of Geography, Macalester College, Saint Paul, USA.

Kelly Biedenweg, Department of Fisheries and Wildlife, Oregon State University, Corvallis, USA.

Austin Charron, Center for Russia, East Europe, and Central Asia, University of Wisconsin–Madison, USA.

Christina E. Dando, Department of Geography and Geology, University of Nebraska Omaha, USA.

Alexander C. Diener, Department of Geography and Atmospheric Science, University of Kansas, Lawrence, USA.

Whitney Fleming, Department of Fisheries and Wildlife, Oregon State University, Corvallis, USA.

Joshua Hagen, College of Letters and Science, University of Wisconsin-Stevens Point, USA.

Andrew Husa, Geography Department, School of Global Integrative Studies, University of Nebraska-Lincoln, USA.

Cheryl E. Morse, Department of Geography and Geosciences, University of Vermont, Burlington, USA.

Amangul Shugatai, Regional Studies Department, International Affairs Institute, Mongolian Academy of Sciences, Ulaanbaatar, Mongolia.

Jeffrey S. Smith, Department of Geography and Geospatial Sciences, Kansas State University, Manhattan, USA.

David J. Trimbach, Washington Department of Fish and Wildlife, USA.

THE POWER OF PLACE IN PLACE ATTACHMENT

ALEXANDER C. DIENER and JOSHUA HAGEN

ABSTRACT. Place attachment is a broad and burgeoning field of scholarship that, while maturing in theory, method, and application, would benefit greatly from more explicit contributions by geographers. Place attachment obviously relates to concepts of residency, including key questions such as: Who moves and why? Who stays and why? Who returns and why? But place attachment also encompasses broader networks of place and geographic contingency, including questions such as: How do place attachments form? Why do people form attachments to some places and not others? How do people form notions of home and homeland within and across varied conditions of mobility? This special issue presents multifaceted theories, methods, and applications inherent in geographical perspectives and suggests that advancing such approaches helps elide reductionism and highlights the power of place in place attachment. *Keywords: place attachment, territorialization, environmental psychology, homeland.*

Place attachment is intrinsic to the human condition and fundamental to our engagement with society and nature. Through processes of spatial organization, demarcation, and differentiation, people develop emotional attachments to specific places, as well as the relationships, objects, practices, and institutions associated with those places. While some suggest that "home"—as a multiscalar concept framing daily life for individuals, families, communities, and societies—is the locus of place attachment, we suggest a broader scope. Rather than fixed upon a specific scale or modality of geographic relationship, place attachment is networked, flexible, and contingent. Accordingly, place attachments manifest in diverse themes, ranging from love, security, and belonging, to departure, loss, and exile, to fear, chauvinism, and avoidance, to isolation, confinement, and cloister, to longing, nostalgia, and return.

This special issue provides a snapshot of the diverse methodological and theoretical perspectives currently animating research on place attachment among geographers. Rather than emphasizing divergent disciplinary approaches, this collection of articles aims to invigorate scholarship on place attachment by demonstrating the complementarity of geography's robust array of place-based theories, methods, and applications. Geography's multifaceted conception of place elides reductionism by highlighting the intersecting influences of nature, society, and meaning upon people-place bonds (Sack 1997; Malpas 2018).

Geographer Tim Cresswell defines "place" as spaces that people have made meaningful and are attached to in some way (2004, 24). In doing so, he suggests that place attachment is an inherent aspect of place and therefore requires deep and textured inquiry regarding its formation and effect. A similar idea is proffered by geographer J. K. Wright (1947), when reminding us that "sense of place" is subjective and therefore *terrae cognitae* only to ourselves. Do individuals and groups need to be attached to place? If the answer is no, then how do they function devoid of place attachment? If the answer is yes, then how can individuals and groups prevent their geographic affections, affinities, and biases from translating to selfishness, exclusivity, and defensiveness? We hope this special issue will promote further consideration of the processes and implications of creating, altering, preserving, and, at times, abandoning and destroying places. As demonstrated by the eclectic topical, methodological, and theoretical approaches taken in the articles comprising this special issue, geographers have much to offer the maturing field of "place-attachment research."

Article Summaries

The articles of this special issue collectively advocate for geographical methods and epistemologies to take a more prominent role in the burgeoning field of place-attachment scholarship. The varied perspectives contained in this special issue are just a small sampling of the greater contribution that geographers might make. Jeffrey S. Smith and Ricardo Aranha lead off the special issue by examining how cognitive or mental mapping can be combined with ethnographic methods to gauge the intensity and texture of place attachments. Smith and Aranha employ this mixed methodology to understand place attachments to Aggieville, a business district in Manhattan, Kansas, that caters to students attending the adjacent Kansas State University. By engaging student groups in participatory visual methods, such as walking interviews, Smith and Aranha reveal that students demonstrated functional, territorial, and emotional attachments to Aggieville. Functional attachments reflected the district's role as a shopping center providing many basic necessities, while territorial attachments found expression though notions of Aggieville as a student-centric space dedicated to their commercial and social practices. Finally, Aggieville engenders emotional attachments as a central venue for student social life, especially nightlife focused on the district's concentration of bars. Beyond the specifics of Aggieville, the article demonstrates the utility of cognitive mapping and participatory visual methods in place-attachment research.

Andrew Husa and Cheryl E. Morse shift focus to examine the connections between place attachment and migration decisions among residents of Nebraska. Like other rural parts of the Great Plains, rural Nebraska has experienced several decades of depopulation due to sustained emigration of residents to urban areas and to other states. Previous research has tended to focus on explaining why people choose to leave rural areas. Differentiating between "leavers" and "stayers," Husa and Morse flip that tendency to examine the converse, or why people made the decision to stay, and how feelings of place attachment and overall contentment correlate to residential decision making. Based on quantitative and qualitative data drawn from an online survey, Husa and Morse find that residents of small towns, defined here as communities with less than 2,500 residents, expressed higher levels of contentment and place attachment compared to respondents from urban areas, and by extension a greater propensity to remain in Nebraska. In contrast to the prevailing tendency to view residents of rural areas experiencing depopulation as somehow "stuck" or "left behind," the article suggests rural residents may have different notions of place attachment influencing their decisions to live in rural areas.

Length of residency also figures prominently in David J. Trimbach, Whitney Fleming, and Kelly Biedenweg's investigation of the connections between place attachment and environmental stewardship in Washington state's Puget Sound region. It is often assumed that length of residency correlates positively with place attachment and concern for environmental issues—that longer-term residents will exhibit greater attachments to their particular place of residency and concern for its environmental issues—while relative newcomers are less likely to manifest those feelings, concerns, and related behaviors. Trimbach, Fleming, and Biedenweg draw from a survey of Puget Sound residents to find that differences in length of residency do not correlate to significant differences in expressions of place attachment or the frequency of engaging in behaviors associated with environmental stewardship, suggesting equal opportunities for engaging longer-term residents and newcomers in addressing ecological sustainability in one of the country's fastest growing regions.

Christina E. Dando's article also explores the intersections of place attachment and environmental concerns. Specifically, Dando employs discourse analysis to examine media coverage of the controversies surrounding the Keystone XL and Dakota Access pipeline projects in the Great Plains. Much of the country imagines the Great Plains as an empty or even "placeless" region, raising questions of why routing pipelines through the region would be controversial. To counter those assumptions, residents of the Great Plains opposed to the pipelines have often framed their positions using notions of place attachment. Even groups often at odds on other issues, such as ranchers and farmers and Native American tribes, have found common ground articulating opposition to the pipelines through notions of place attachments to the Great Plains region as supporting their homes, livelihoods, and identities.

Reflecting the influence of psychology and sociology, place-attachment research has tended to focus on Western societies, but Austin Charron's study of people displaced to mainland Ukraine from Crimea following Russia's annexation of that region in 2014 illustrates how well positioned geographers are to expand the scope of place-attachment research. Crimea is populated primarily by ethnic Russians, Ukrainians, and Crimean Tatars, with the latter group representing the indigenous population. Subject to various forms of persecution and repression culminating in decades of exile from Crimea during the Soviet period, Crimean Tatars constituted a relatively small minority even before Russia's annexation of the region displaced many to mainland Ukraine. Sizable numbers of ethnic Ukrainians were also displaced. Drawing on ethnographic fieldwork among those displaced populations, Charron investigates how competing historical narratives and practices inform senses of place attachment to Crimea as an ethnic homeland, even under conditions of repeated and prolonged displacement and exile.

In a related vein, Holly R. Barcus and Amangul Shugatai study the role of nostalgia as a process of selective remembering to communicate shared notions of place attachment across generations among diasporic groups. Ethnic Kazakhs are a substantial minority across much of Mongolia and even constitute a majority in parts of western Mongolia. Since the fall of communism in 1990, many ethnic Kazakhs from western Mongolia have moved to larger urban centers, especially the capital city Ulaanbaatar, to escape the poverty and severe environmental conditions that characterize much of life in that region. Despite those harsh realities, Barcus and Shugatai show how the Kazakh diasporic community across Mongolia increasingly remembers western Mongolia as a bucolic steppe landscape intimately tied to traditional Kazakh culture and community. This romanticization of life on the western Mongolian steppe highlights how nostalgia operates as a multigenerational process to foster notions of place attachment and ethnic belonging among diasporic communities.

Northern Kazakhstan bears some similarities to western Mongolia in that it also contains sizable ethnic minority populations, most notably populations of ethnic Russians who settled across the region throughout the Czarist and Soviet periods. Alexander C. Diener focuses on this multiethnic borderland and persistent concerns that its ethnic Russian population could serve as justification for secessionist demands or annexation by Russia, not unlike developments in eastern Ukraine. Despite its substantial ethnic Russian population, Diener finds that calls for independence or unification with Russia are not frequent and do not appear to enjoy substantial support across Kazakhstan's northern borderlands. This study also highlights how place-attachment scholarship tends to focus on the scales of home and locality or state and homeland while ignoring how regional or borderland scales illustrate the mutable and multifaceted nature of place identities and place attachments.

Given the prevalence of psychology and sociology and ethnographic methods, it is hardly surprising that place-attachment scholarship has tended to focus on contemporary cases. Joshua Hagen's examination of historic preservation and places of memory highlights the possibility of adding historical texture to our understanding of place attachment. Reflecting its totalitarian impulses, the Nazi regime sponsored a variety of building projects conveying its interpretation of national identity, history, and homeland. In that sense, the regime deployed historic preservation as a tool to create places of memory and place attachments supporting its ideology and authority. Cumulatively, these places of memory exemplified the racial ideology of blood and soil that animated Nazi ideology.

To conclude this special issue, Alexander C. Diener and Joshua Hagen sketch out a place-based model that positions place attachment as an assemblage of nature, meaning, and social relations. Such a framework emphasizes the contingent and varied dynamics of human agency, including human-environment interaction and individual-collective identities, to make and remake places and place attachments. Place attachment has always been a fundamental concern within geography, but since the 1980s, geographers have tended to engage the topic on a more implicit basis. As a result, environmental psychology and sociology came to dominate place-attachment research. Through a more holistic and explicit engagement with the geographies of place attachment, Diener and Hagen offer a place-based model that promotes understanding of the territorial processes undergirding place attachment and the possibilities for reconciling local belonging and cosmopolitan values.

Cumulatively, this special issue highlights the potential for geography as a discipline to assume a higher profile in the field of place-attachment scholarship. Scholars from a variety of other disciplines have made important strides in advancing our understanding of the varied dynamics of place attachment, but those disciplinary perspectives have tended to minimize geographical differences, often assuming space and place to be little more than relative location. In contrast, sustained and explicit engagement on the part of geographers holds great promise for broadening and deepening our appreciation for the power of place in place attachment.

Acknowledgments

As guest editors of this special issue, Alexander and Joshua thank the contributors for sharing their varied expertise and perspectives on place attachment, as well as the anonymous reviewers for providing insightful comments. We also thank Dave Kaplan for his guidance as the special issue took shape and proceeded toward publication.

Funding

This work was supported by the University of Kansas [General Research Fund].

References

Cresswell, T. 2004. *Place: An Introduction*. Malden, Mass: Blackwell.

Malpas, J. 2018. *Place and Experience: A Philosophical Topography*. New York: Routledge.

Sack, R. D. 1997. *Home Geographicus: A Framework for Action, Awareness, and Moral Concern*. Baltimore, Md.: Johns Hopkins University Press.

Wright, J. K. 1947. Terrae Incognitae: The Place of the Imagination in Geography. *Annals of the Association of American Geographers* 37 (1): 1–15.

COGNITIVE MAPPING AS A METHOD TO ASSESS PEOPLES' ATTACHMENT TO PLACE

JEFFREY S. SMITH and RICARDO ARANHA

ABSTRACT. Over the past five decades, the methodological approaches used to understand peoples' emotional ties to place have been incredibly diverse, employing both quantitative and qualitative techniques (Lewicka 2011). In recent years more ethnographic, participatory visual methods such as cognitive (mental) mapping, narrative mapping, and psychogeography are showing considerable promise in revealing peoples' attachment to place. This article seeks to illustrate how mental mapping and walking interviews can be used as an effective tool in exposing peoples' deep emotional connection to place. With a focus on a student-centered business district in Manhattan, Kansas, we demonstrate that Aggieville exemplifies all three types of emotional attachment: functional, territorial, and emotional. Students are functionally attached to Aggieville because it provides nearly all their foundational needs (e.g. food, clothing, housing, employment) enabling them to achieve their life goals. Students demonstrate territorial attachment because Aggieville is a space created by students for students and the accepted activities and behavior are intended for students only. Students have also established an emotional attachment to "The Ville" because it is an important place for them to socialize and bond with others. *Keywords: Aggieville, cognitive mapping, mental maps, place attachment, psychogeography, walking interviews.*

"To be rooted [in place] is perhaps the most important and least recognized need of the human soul. … A human being has roots by virtue of his real, active, and natural participation in the life of a community, which preserves in living shape certain treasures of the past and certain expectations for the future. This participation is a natural one, brought about by place, conditions of birth, profession, and social surroundings." – Simone Weil

Simone Weil's quote reminds us that place matters and our interaction with place is a fundamental element of our lives. As individuals, and members of a-community, much of our identity is derived from the connections we make with places. Some places affect us more profoundly than others and, for a multitude of reasons, we attach greater significance and develop deeper emotional connections to those places.

Among the world's Spanish-speaking population, the term *querencia* (the place of your heart's desire) captures the essence of this special feeling people develop for certain places. The term refers to the place that resonates with your soul; the place where you feel most comfortable and content (Fountleroy 1997). The same concept is found in countless other cultures throughout the world. In the German language, for example, the word *heimat* defies easy translation into

English but encapsulates the ideas of homeland as well as tradition, belonging, identity, landscape, and place (Economist 2018).

A sizeable and growing body of literature refers to the strong feelings that people develop for certain places as place attachment. Other terms include love of place (Tuan 1974), place bonding (Hernández and others 2014; Jorgensen and Stedman 2006), rootedness (Tuan 1980), insideness (Relph 1976; Rowles 1990; Seamon 2008), and emotional ties to place (Chamlee-Wright and Storr 2009). The common denominator is that these are places people feel so passionate about that much of their identity is derived from their interaction with that place.

Since the 1960s, scholars across numerous disciplines have applied various methods and techniques to increase our understanding of peoples' attachment to place. As Lynn Manzo and Patrick Devine-Wright (2014) explain, the approaches taken have been as diverse as the scholars conducting the research. There have been some common trends, however. Environmental psychologists, sociologists, and community planners have been partial to quantitative methods that measure peoples' varying degrees of emotional connection to a place. Their research commonly asks individuals to indicate their affinity for a particular place using a Likert scale. Among these scholars, little attention is given to the qualities of place. By comparison, scholars in such fields as art history, cultural geography, and regional studies have employed more qualitative methods. Their objective has been to understand the nuances of place and the deep-seated feelings people have for them as well as to expose how people interact with a place and the meaning they hold for that place. These goals are achieved through a variety of means, but most can be distilled down to ethnographic approaches.

In recent years there has been a growing interest in using cognitive mapping (also referred to as mental mapping, sketch mapping, participatory mapping, narrative mapping, pictorial mapping, and memory work) to assess peoples' emotional connection to place (Tsai and Lo 2013; Boschmann and Cubbon 2014; Curtis 2016). Regardless of the lexicon, this technique is grounded in the field of phenomenology, which seeks to expose how people develop a relationship with a particular place (Seamon 2018). The ultimate goal is to unmask, reveal, and interpret how individuals interact with, and become attached to, a place. Yet to be fully explored, however, is how mental mapping techniques can be used to understand peoples' attachment to place. The purpose of this article is to help fill that gap in the literature by showcasing how cognitive (mental) mapping can be used as an effective tool in capturing the character of place and exposing peoples' deep emotional connection to place.

Advances in Cognitive Mapping

The Chicago School of Urban Studies has had a profound and lasting impact on scholarship far beyond its primary scope of understanding the spatial organization of cities. Of particular note is that the wide-ranging research that emerged from the shores of Lake Michigan has added to our understanding of the character of place.

For example, in his book *The Image of the City* (1960), Chicago-native Kevin Lynch introduces an innovative approach to comprehending the perceptions people hold for urban environments. He explains that as people navigate (wayfind) through their environs, they attach meaning and significance to certain places within the built environment. Asking people to sketch their mental (cognitive) map of a place can help extract information that is otherwise difficult for them to articulate (Endnote 1). One of the main outcomes of Lynch's work is that peoples' mental maps typically include five key elements. *Edges* are the boundaries or outer limits of a mental map. *Pathways* are the routes of frequent travel, the threads that hold the map together. *Nodes* are junction places—locations where pathways come together. *Districts* are areas with a common identity, while *landmarks* are reference points that stand out in peoples' minds.

At the same time, Lynch was advancing the utility of mental maps, Guy Debord and the Paris School of Situationists was exploring the personality of place. With the intent of deconstructing the everyday scene, the Situationists' goal was to help individuals recognize the relationship they have with the built environment around them. According to Debord, the best way to achieve this is *à pied* (by foot). The popular nineteenth-century French word *flâneur* undoubtedly resonated with Debord because of its reference to an educated man of leisure wandering about and critically observing the urban scene. This captures the essence of Paris School's main intent—to understand how place affects peoples' emotions and behavior.

An intriguing example of the Situationists School comes from Will Self and Ralph Steadman (2007), who teamed up to explore the sundry of emotions people have for places around the world. Their book *Psychogeography*, written and illustrated in the aftermath of September 11, 2001, encourages readers to walk through their community and be aware of how the built environment elicits various emotional feelings. For example, does a mosque elicit feelings of peace and spirituality or fear and terrorism? As disciples of Debord, Self and Steadman assert that society can use the principles of psychogeography to elevate their awareness of how their emotions intersect with their surroundings.

Mental (cognitive) mapping, wayfinding, walking interview, narrative mapping, and psychogeography are among the many forms of outdoor-based methods that all fall under the larger umbrella of participatory visual methods (Kusenbach 2003; Ingold and Lee 2008; Evan and Jones 2011; Hanna and others 2018). This approach to ethnographic research seeks to combine the intuition and insight of the researcher with the knowledge and experience of the local participant. Two benefits of this methodological approach include: helping break down the distance between etic and emic voices and enabling the researcher and informant to work as equals; and enabling scholars to explore more nuanced narratives that lie at the heart of place attachment. Critics of the method assert, however, that each nugget of information collected is of limited value (*a la* anecdote) because it has its own site, time, and human experience (Rishbeth

2014). This is a concern especially among those who seek to follow objective, standardized procedures in their research methods. Michael Patterson and Daniel Williams (2005) counter this argument by saying that place attachment research would do well to use a wider range of people-place methods to gain insight into the deep emotional connection people have for place.

Approaches to Place Attachment Research

Research on understanding peoples' emotional connection to place is vast and ever expanding. However, as Leila Scannell and Robert Gifford (2010) explain in their tripartite model, most research boils down to three dimensions. Social pychologists, sociologists, and anthropologists have focused on the *person* component. This approach seeks to understand *who* (both individuals and groups of people) is attached to a place. By comparison, geographers have led the charge in exploring the *place* component. This line of research seeks to expose the characteristics of *what* people are attached to across various scales. The final component (*process*) has been primarily the domain of environmental psychologists. Their objective is to understand and articulate *how* a place becomes meaningful and *how* the mind reacts to being in a place. Psychologists have determined that length of time in residence is one of the strongest predictors of peoples' feelings of place attachment.

After conducting a thorough review of the field, Maria Lewicka (2011) found that the person and process components have received the most attention, while place has received much less scrutiny. Moreover, scholars have placed greater emphasis on understanding differences among individuals at the expense of elucidating the characteristics of place. In Lewicka's estimation, this skewed research perspective has retarded the overall development of the field. Clearly, more research is needed on understanding the qualities of place.

An edited volume titled *Explorations in Place Attachment* (Smith 2018) has helped fill that gap in the literature by identifying six types of places to which people become emotionally attached. The collection of essays defines and explains *secure places* (locations where people feel emotionally secure), *socializing places, transformative places* (where significant events unfolded in a person's lifetime), *restorative places* (locations that foster physical, emotional, and intellectual health), *validating places* (settings where human activity is infused with shared meaning), and *vanishing places* (locations that have been lost or are at risk). Of particular relevance to this article are socializing places.

According to the work of numerous geographers, one of the most effective ways in which people develop a strong attachment to place is through the interaction they have with others (Milligan 1998; Barcus and Brunn 2009; Smith and Cartlidge 2011). Moreover, these relationships typically unfold in specific locations (Massey 2005). Socializing places are venues that become particularly meaningful because they afford individuals the opportunity to connect with others. They are locations with a strong sense of community; places where people feel welcome.

Ray Oldenburg's research on *Third Places* is a prime example. As Oldenburg (1989) explains, peoples' *primary place* is their home, while their *secondary place* is their place of employment. These are the locations where people typically spend most of their time during the week. *Third places* are locations where people socialize, including cafes, night clubs, barber shops, community centers, and houses of worship. In their book *Scenescapes*, Daniel Silver and Terry Clark (2016) testify that socializing places are complex and chockful of hidden messages. However, their book helps readers decipher the cryptic narratives and see which social activities are tied to a certain location.

In an exhaustive study, Blake Gumprecht (2008) declares that two of the ten defining characteristics of America's college towns are their large youthful population and their tolerance for unconventional behavior. Youthful refers not only to the disproportionate number of twenty-something year olds in town, but also to the uninhibited, free-spirited lifestyle that goes hand in hand with tolerance for eccentric and outright indulgent behavior. Although located in a decidedly conservative state, Manhattan, Kansas, is a quintessential college town. Moreover, one commercial area in particular is a haven for youthful behavior with surprising tolerance for excessive, unconventional behavior. Aggieville is an example of both a socializing place and a place to which people have developed a strong emotional attachment.

Aggieville: A Place to Socialize

Located kitty-corner from the southeast corner of Kansas State University's main campus, "The Ville" holds a special place in the hearts of countless students, past and present (Figure 1). Yet, Aggieville had a rather peculiar beginning. Like other land-grant colleges, when Kansas State Agricultural College was founded in 1863, it was located a considerable distance from downtown. This remote, rural location was ideal for students studying agricultural practices and animal husbandry, but was terribly inconvenient when they needed goods and services.

In the early years, students had to walk three miles (one-way) to purchase their school supplies and groceries. Moreover, because the first dormitory was not built until 1926, nearby homeowners (professors) offered room and board to students at inflated prices. Although college administrators made various attempts to open a campus bookstore, none were successful. Fed up with the high cost of their education, in 1899 students established a nearby co-op bookstore, dining hall, laundry, and barbershop that met many of their needs at more reasonable prices. Ironically, if college administrators had successfully run a bookstore, Aggieville might not have developed in the manner it did (Gumprecht 2008).

The Student Co-Operative Book Store became the focal point around which Aggieville grew. Over the next decade, other key businesses opened including

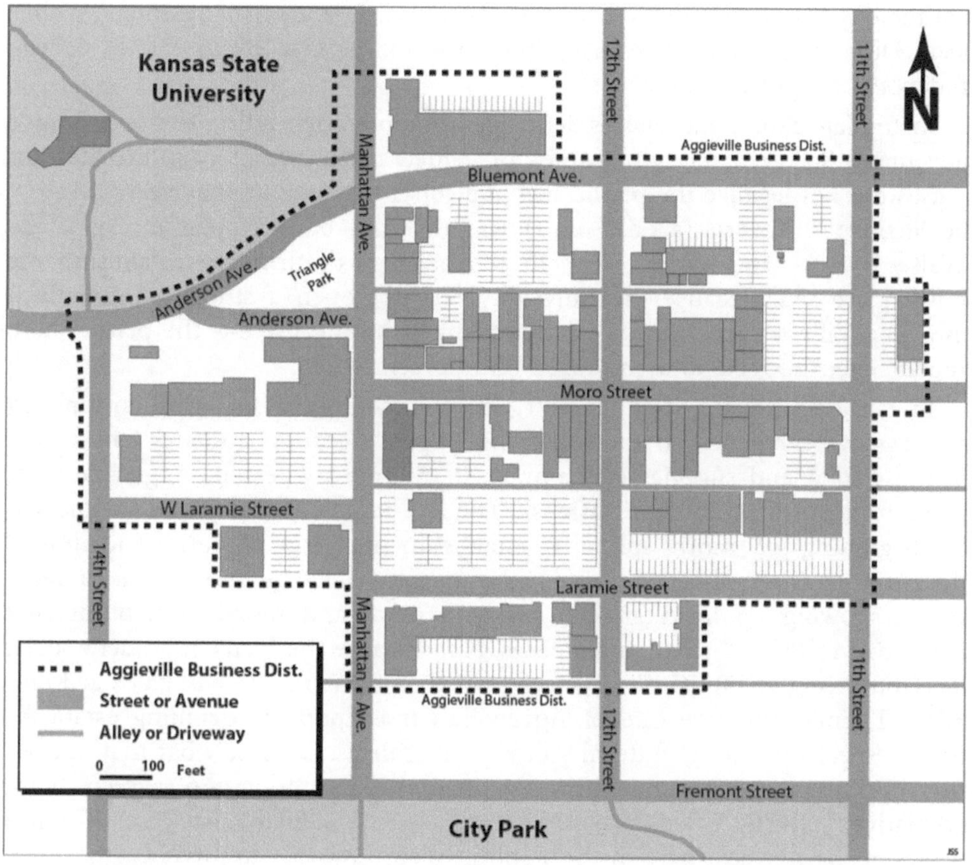

FIG. 1—Map of Aggieville Business District. Cartography by Jeffrey S. Smith.

a grocery, another bookstore, men's clothing, and a restaurant. By 1909, Manhattan city officials finally began offering streetcar service between Aggieville and downtown. This may have offered students greater options in where they could purchase their necessities, but most still preferred shopping in Aggieville. In 1910, the city of Manhattan annexed Aggieville and homeowners began moving out (Walter 1989).

Over time, each private home vacated was replaced with a new business. The organic nature of Aggieville ensured that most businesses catered to students' needs and wants. According to the local newspaper, by 1913 Aggieville hosted twenty-three student-centered businesses, including a bookstore, restaurant, laundry, grocery, and barbershop, as well as a bakery, sporting-goods store, bank, post office, and supply store for cooking and sewing classes. In 1915, Aggieville took on a new vibe when the grocery store owner opened a dance hall on the building's second floor. By the 1940s, crowds had become so large that the city fire marshal was forced to restrict attendance to 750 people (Walter 1989). Despite all the changes, Aggieville was still defined by the life of college students (Gumprecht 2008).

In the late 1800s, Kansas residents were famous for leading the temperance movement. So, even though Prohibition ended nationally in 1933, the State of Kansas did not legalize the public sale of alcohol for another four years. In 1937, the Shamrock Tavern (a.k.a. Slim's) was the first bar to open in Aggieville (Walter 1989). After World War II, the number of drinking establishments slowly increased. Because the university did not open a student union until 1956, The Ville became the de facto student gathering place—-the place where students ate, shopped, and socialized (Gumprecht 2008).

In 1956, the Kansas State Student Union, including a dining room, opened and Aggieville business owners felt the impact because students could now purchase food and supplies on campus. Restaurant owners in Aggieville were forced to adapt or go out of business. The late 1950s and early 1960s witnessed the beginning of more specialty establishments (e.g. pizzeria and pubs). Moreover, by the mid-1960s, social mores began to change and female students started showing up in Aggieville bars. As more and more women inserted themselves in these former male spaces, bars became the focus of student social life (Gumprecht 2008). Kite's (a tavern owned by former MLB baseball star Keith "Kite" Thomas) became one of Aggieville's most popular drinking establishments, especially among 'fraternity boys' and athletes. Each new bar that opened catered to a particular niche within the student body population (e.g. cowboys, alternative, LGBTQIA+). In 1997, *Playboy* magazine named Rusty's Last Chance as one of the ten best bars in the nation. According to Gumprecht, "the most enduring impact of the 1960s on Aggieville was an increase in the number and size of bars and the growing importance of the bar scene to student life. … Aggieville became known as a bar district" (2008, 127).

Beginning in the 1970s, drinking ceased to become a weekend-only activity. Each week, starting on Thursday afternoons and ending during the wee hours of Sunday morning, crowds of students descended upon Aggieville. As one report indicated, by 4:00 pm on most Fridays (during the school year) the wait time to get a barstool in Aggieville is impossibly long (Walter 1989). In the 1980s, barhopping became a popular social event among members of Greek houses and other student groups. In 2007, Aggieville bar owners introduced Fake Patty's Day celebration. Because St. Patrick's Day usually falls during Spring Break, the owners hold the event on the closest Saturday to March 17. The annual festival draws thousands of young adults from throughout the region and has become so popular that state police are temporarily assigned to help with crowd control and public drunkenness. It has become one of the most anticipated social events of the year.

In recent years, Aggieville has fewer and fewer retail stores and an increasing number of drinking establishments. At last count, there are twenty-five bars, not including businesses with liquor licenses whose main focus is selling food (ABA 2016). Without question, Aggieville is an excellent example of a socializing place (Figure 2). At the same time, for various reasons, students past and present hold a deep affinity for Aggieville. As one unidentified student once declared, "I'm double majoring in Business and Aggieville."

Methods—Participatory Visual Methods

With the purpose of illustrating how cognitive (mental) mapping can be used as an effective tool in understanding character of place and peoples' deep emotional attachment to place, data were derived from mental maps produced by students at Kansas State University. Over the course of his career, the lead author has taught more than 10,000 students in classes ranging from introductory-level World and Human Geography to graduate-level seminars. In most of those classes, students (from diverse academic backgrounds) were asked to draw their mental map of Manhattan, Kansas, with as much detail as possible. Using students' cognitive maps has proven abundantly effective in reinforcing various lecture topics. In 2011, he began digitizing students' maps. Serendipitously, a large data base (n=483) of cognitive maps were retained in digital format. Each of the 483 mental maps were perused for evidence of emotional connections toward Aggieville.

Fig. 2—Photo of the core of Aggieville. Image shows a good number of bars lining Moro Street. Photo by Jeffrey S. Smith, December 2019.

Additionally, in November 2019, 12 members of the K-State Gamma Theta Upsilon Geography Club (undergraduate and graduate students) volunteered to participate in psychogeography and a walking interview through Aggieville. Because the core of Aggieville (excluding parking lots) covers only about five square blocks, it is possible to walk the entire district in an afternoon. Each student was given a blank street map of Aggieville and asked to mark their emotional response (using highlighters – green [positive], yellow [neutral], and red [negative]) for locations along the walk through the business district. They were also asked to write detailed notes in the margins. These maps provided greater detail on a student's emotional response to place than the mental maps normally permit. The remainder of this article sets the stage with a brief overview of the three categories of emotional attachment to place. Then, it articulates how cognitive mapping techniques can be used to capture the character of place and understand peoples' emotional attachment to place.

Understanding Students Attachment to "The Ville"

Existing literature indicates that there are three types of emotional attachment to place (Livingston and others 2008). One is *functional (a.k.a. practical) attachment*, which refers to the capacity of a place to meet peoples' needs for survival and life goals. This type of attachment centers on a place's ability to sustain life and enable an individual to grow both physically and intellectually. Feelings of functional attachment occur when a place allows a person to mature, pursue individual happiness, and reach their potential.

The second type of attachment is *territorial attachment*, which centers on the connection people have for a space they control. Individuals or groups control space through possession, personalization, and regulation. From an early age, individuals become attached to the spaces they frequently occupy (e.g. baby crib, bedroom, house, neighborhood) (Morgan 2010). One way we exercise control over places is by establishing ownership or possession (formally or informally) of that space. We seek to buy it or gain lawful title to it. When ownership of a space is not possible, then we seek to control it by personalizing it. This involves covering the space with our personal belongings. This tells others that this place is ours. We also demonstrate territorial attachment by regulating who is permitted to use or occupy a particular space. This is achieved by determining the type of people who occupy a space or the appropriate activities that unfold within a given space (Mitchell 2003). Through the acts of possession, personalization, and/or regulation, we establish territorial control and attachment to particular places.

The third type of attachment is *emotional attachment*. This is the deep-seated feelings people develop for a place. Preliminary research suggests that certain places become meaningful because they are associated with cycles of emotional triggers that repeatedly arouse and stimulate an emotional response (Morgan 2010). These emotional responses become so ingrained in our psyche that we develop emotional attachment to that place. Edward Relph's (1976) work on *insideness* is at the core of emotional attachment. People feel strong emotional connections to places that are intimate and familiar.

When examining the 483 mental maps that students from Kansas State University have produced over the past eight years, certain places within the city of Manhattan consistently stand out, including Aggieville (78 percent), Walmart and/or Target (58 percent), and Bill Snyder Family (football) Stadium (56 percent) (Table 1). Even students who are geographically challenged and have underdeveloped mental maps typically include Aggieville on their sketch map (Figure 3). It is plainly evident that Aggieville holds a special place in the hearts and minds of K-State students. For many students, Aggieville sits near or at the bull's eye of their mental map and is a central place in their lives (Figure 4). The way in which students interact with the business district demonstrates that Aggieville is an excellent example of a place that encompasses all three types of attachment: functional, territorial, and emotional.

TABLE 1—FREQUENCY OF PLACES (DISTRICTS AND LANDMARKS) APPEARING ON STUDENTS' MENTAL MAPS

PLACES IDENTIFIED	PERCENT OF MAPS INCLUDING PLACE
Aggieville	78 percent
Walmart and/or Target	58 percent
Bill Snyder Family (Football) Stadium	56 percent
Shopping Mall / Big Box Stores	48 percent
Manhattan Hill / Tuttle Creek Reservoir	31 percent
Downtown / C.B.D.	28 percent
Fast Food (McDonalds, Sonic)	24 percent
City Park(s)	22 percent
Manhattan K-12 Schools	19 percent
Agricultural lands / Cows / Barns	16 percent
Fraternity / Sorority House(s)	14 percent
Anderson Hall (Univ. Admin.)	12 percent
Theater / Cinema	12 percent
Church(es)	10 percent
Student Union	10 percent
Pizza Restaurant(s)	9 percent
Grocery Store(s)	7 percent
Hale (University) Library	7 percent
Bank(s) / ATM Machines	5 percent
Police Station	5 percent
Zoo	5 percent
Hospital	3 percent
K-StatePowercatAthleticIcon	9 percent

Source: Mental maps drawn by students at Kansas State University between 2011 and 2019 (n=483).

FIG. 3—A student's underdeveloped mental map of Manhattan, Kansas. Despite not having a clue, Aggieville, K-State campus, and the football stadium are all identified.

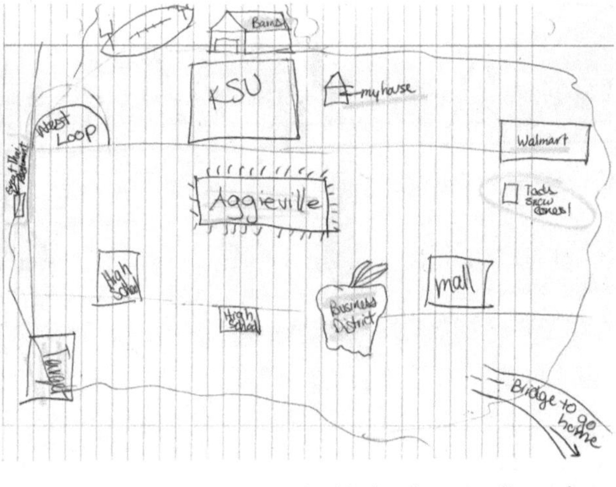

FIG. 4—A student's mental map showing a highlighted Aggieville at the center of their map.

From its very beginning, Aggieville has been a place created by students, for students. Today, Aggieville continues to serve that important function. Students' mental maps indicate that Aggieville is one of the key locations within Manhattan where they can go to eat their meals, find employment, and purchase essential things for their lives (school supplies, books, computer and phone repairs, coffee and snacks, clothing, souvenirs, and gifts, as well as a laundromat and eye doctor) (Figure 5). Although the ensemble of retail shops constantly changes because of high turnover (68 percent of retail businesses turned over between 2011 and 2016) (ABA 2016), Aggieville is still the "go-to" place for college-aged students to get the things they need.

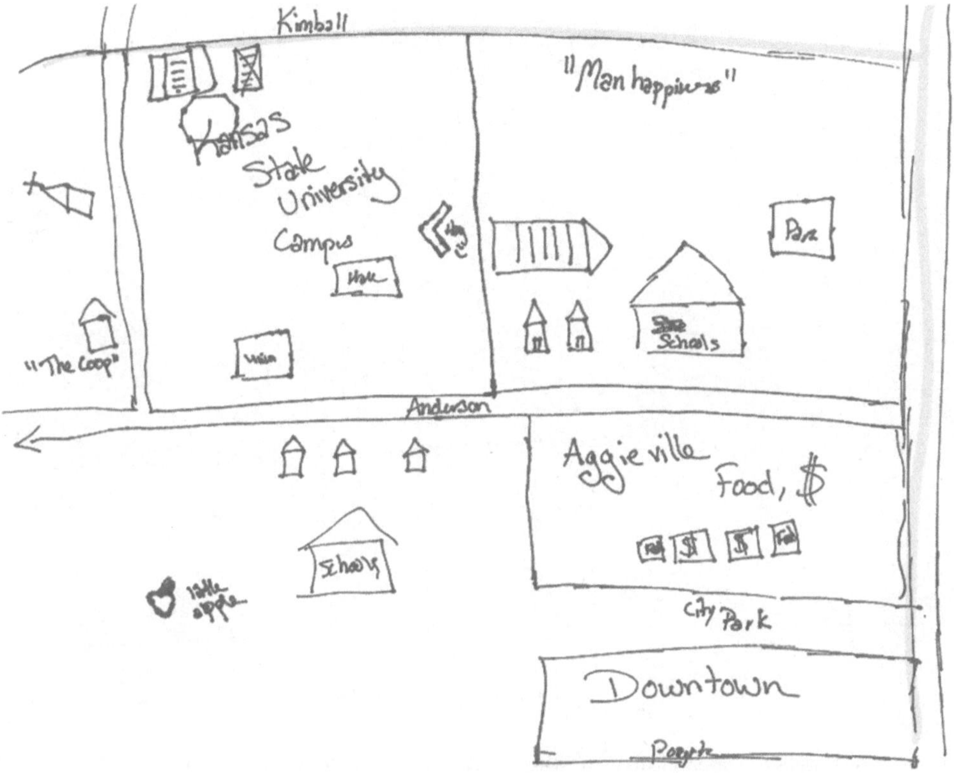

FIG. 5—A student's mental map showing the importance of Aggieville as a source of food and costly entertainment or place of employment.

Often overlooked is the abundant residential spaces (apartments) that are available for rent in Aggieville. According to the Aggieville Business Association (ABA), there are 335 housing units within Aggieville, with approximately 2,000 people living in and around the immediate Aggieville area (ABA 2016). Some students have indicated that, in a typical week, their activity space extends no farther than the K-State campus and Aggieville. Aggieville provides most, if not all, of the essential elements (e.g. food, supplies, clothing, and employment) that students need at this stage in their life (Figure 6). It is abundantly clear that students have a functional attachment to Aggieville.

During our psychogeographical walking tour of Aggieville, students shared stories about how Aggieville is instrumental in their development as individuals. Many of the narratives center on socializing and interacting with peers. One student, however, talked about how his life was transformed when he went to a bar for the first time with his mom and dad. That day is burned into his memory because it was the first time he felt like an adult around his parents. Some places within Aggieville are meaningful because they create a relaxed,

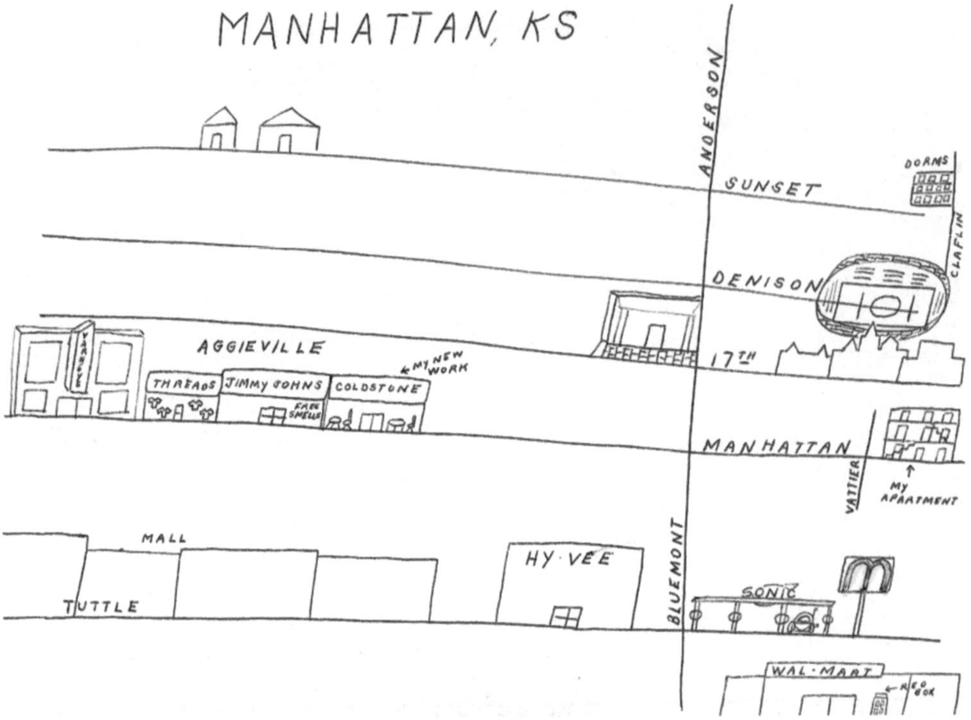

FIG. 6—A student's mental map showing Aggieville in the center of the map. Notice some of the key businesses, including Varney's bookstore, clothing store (Threads), restaurants, and place of employment.

comfortable environment where people can express their individual identity. For example, two students from the walking interview noted that one bar is popular and meaningful among LGBTQIA+ students because they felt safe and secure in that space. Clearly, students feel attached to Aggieville because it helps meet their everyday needs and enables them to grow as individuals.

For a vast majority of Manhattan residents, especially those who are beyond their college years, the activities that dominate Aggieville, particularly during Fake Patty's Day weekend, are nothing less than decadent and debaucherous (Figure 7). Moreover, it is common knowledge that families with young children typically avoid Aggieville on Friday and Saturday nights because the unabashed, lascivious behavior is not appropriate for all eyes (ABA 2016). Yet, this is exactly what appeals to the college-aged crowds that fill the bars. As their mental maps demonstrate, students control Aggieville through a variety of means. For example, with their wallets they cast the decisive vote in which types of businesses are successful. But, the most effective way students mark their territory and control the space is through their promiscuous and self-indulgent behavior (Figure 8). The uninhibited, free-spirited lifestyle that goes hand in hand with outright

FIG. 7—A student's mental map showing the bars of Aggieville surrounded by Greek houses. The bubble quotes read: "Being drunk is really cool" and "Yay! I'm drunk".

imprudent behavior creates a space with a unique character (Gumprecht 2008). That identity does not appeal to all segments of the local population, and so individuals who do not appreciate or embrace such activity are kept at bay.

As we walked through Aggieville during our psychogeographical walking tour, students shared story after story about the wonderful times they have had. Most of those fond memories centered around drinking and socializing to all hours of the night. For example, one student told us about a common birthday celebration called "running the gauntlet." He explained that the practice involves a group of friends who visit nearly every bar in Aggieville, buying the house specialty drink for the birthday person and themselves (Figure 9). According to the students, this can be quite expensive ($60 or more per person) and commonly involves urinating in public and vomiting. Without saying it, the students' stories demonstrated what activities are deemed appropriate for Aggieville. Even city officials acknowledge that Aggieville caters to the student-aged population in town and no matter what measures they take, the city has a difficult time

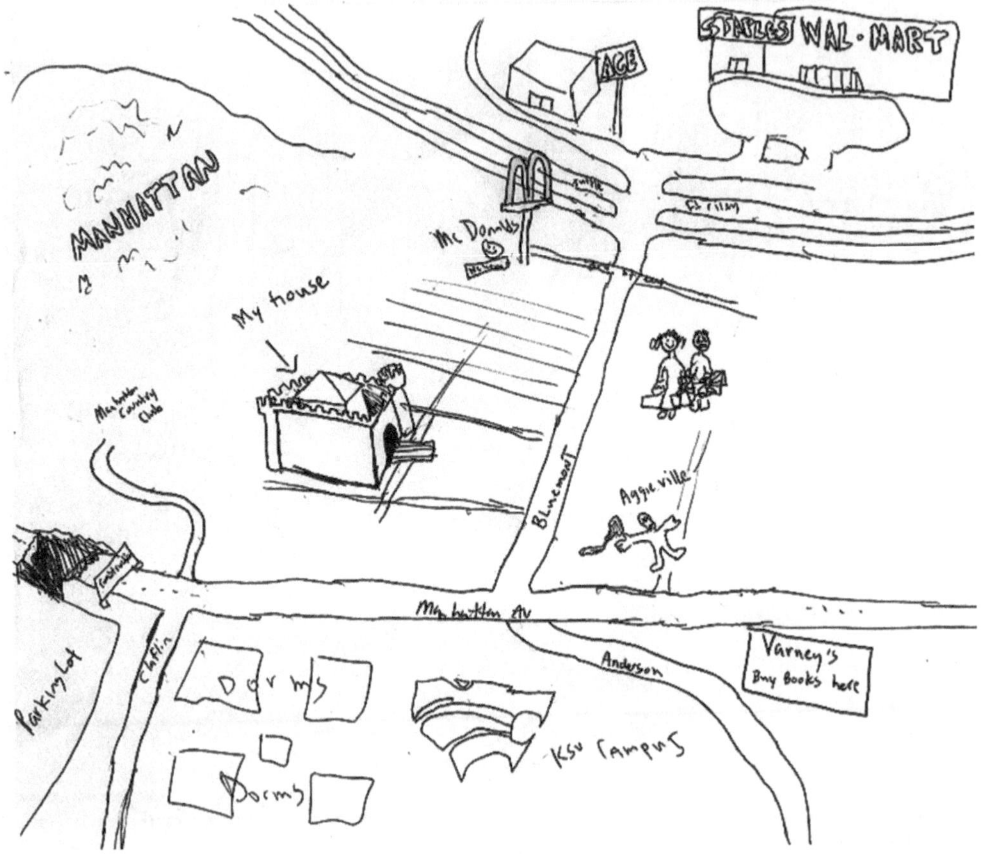

FIG. 8—A student's mental map showing key locations within Manhattan, Kansas. Notice in the vicinity of Aggieville there is a passed-out student with spilled beer next to him and a young couple shopping.

attracting businesses that cater to a spectrum of the population that is outside college age. Those businesses are rarely successful in Aggieville (ABA 2016).

Seventeen short miles west of Manhattan is Fort Riley military base. When awarded leave (e.g. weekends), many of the young, college-age soldiers visit the bars and eateries in Aggieville. This is an economic boon for business owners, especially during the summer months. However, one of the biggest problems city officials and local police face during the academic year is conflicts between students and military personnel. During our walking interview, three students identified the parking lots on the south side of Aggieville as a negative emotional space for a variety of reasons, but a big factor is that fights between students and soldiers commonly break out in these parking lots. Nearly all of the participants agreed that the student-body population and the soldiers mix like oil and water, especially when alcohol and excessive bravado are added to the mix. But more to

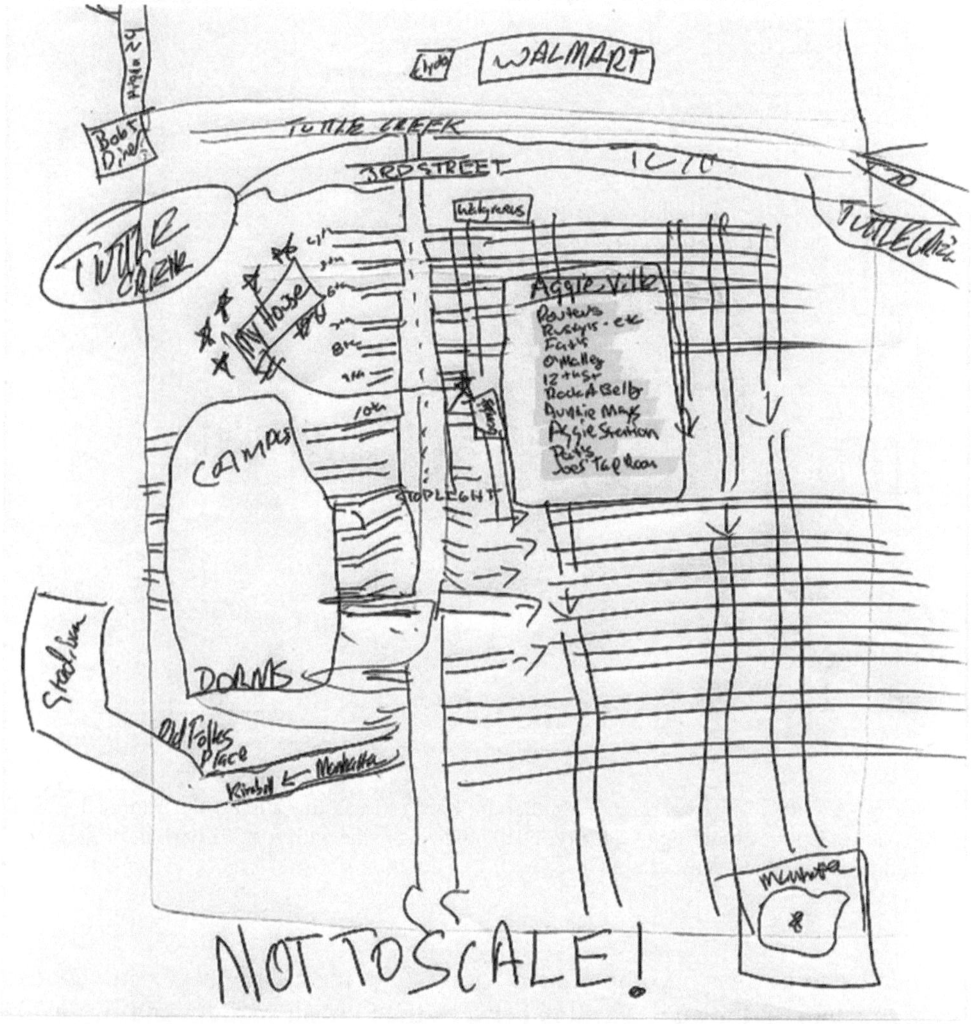

FIG. 9—A student's mental map of Manhattan, Kansas, not drawn to scale. In the area labeled Aggieville is a list of popular bars.

the point is that one segment of the Aggieville customers—college students—regard Aggieville as their own and are willing to defend that space against an outside aggressor (military personnel).

Among past and present students, there are two places within Manhattan, Kansas, that hold deep, unrivaled emotionally meaning: Kansas State University and Aggieville (Figure 10). On the K-State campus, it would be difficult to identify a more emotion laden location than Bill Snyder Family Stadium. Soon after Coach Snyder arrived in 1989, the football field became the venue for significant upsets against formidable visiting opponents. Over the years, the

FIG. 10—A student's mental map of Manhattan, Kansas, including Bill Snyder Family Stadium, the KSU administrative building (Anderson Hall), and Aggieville with specific bars and restaurants identified along Moro Street.

expanded stadium has become a main source of pride as the team's success has put Manhattan on the map. The link between the football stadium and Aggieville is not to be overlooked. Countless times after unexpected or significant victories, Wildcat football fans have rushed the field and torn down the goal posts carrying them the two miles to Aggieville where they are deposited at the entrance of a particularly popular bar. Together, Bill Synder Family Stadium and Aggieville serve as validating places—places where human activity is infused with shared meaning, places where personal and group identity is reinforced (Smith 2018). By participating and investing themselves in a place, participants come to feel a part of that place and associate their identity with that place (Seamon 2014). The stadium, and by extension Aggieville, is a place (especially after winning) where people feel that what is important to them and the things they value have meaning; their lives are validated.

Yet, the emotional significance of Aggieville goes well beyond its connection to sporting victories. As illustrated on an abundance of students' mental maps,

Aggieville is steeped in emotion. In the process of binge drinking, generations of K-State students have engaged in social activities, networked with peers, created lifelong bonds, and even found life partners in Aggieville. Aggieville is at the center of K-State students' social life, making it a quintessential example of a Third Place (Oldenberg 1989).

As we walked through Aggieville with students, the emotional significance of the business district became plainly evident. Students regaled stories about where they had their first beer, spent their twenty-first birthday, and met their partner/spouse. But, it was interesting to hear stories that were not all filled with happy emotions. Some students told anecdotes about being arrested for Minor in Possession (MIP), disorderly conduct violations, fractured bones, and getting mugged or beaten up. Yet, despite these seemingly negative experiences, all of the students—without exception—hold deep-seated feelings of place attachment for Aggieville. Both the cognitive mapping exercise and the psychogeographical walking tour captures this deep-seated connection to place.

Conclusions

Sixty years ago, Kevin Lynch's research introduced scholars to the utility of mental maps in understanding urban structure. In the decades since Lynch's research, a variety of participatory visual methods including cognitive mapping, narrative mapping, walking interviews, and psychogeography are proving effective in capturing the character of place. Mental maps are an excellent way to identify what places are particularly meaningful for individuals. Furthermore, the technique is effective in gathering information about a place that is otherwise difficult for people to articulate. This approach is also showing considerable promise in exposing peoples' attachment to place because of its ability to identify what specific locales are most important to individuals (e.g. Smith and Cartlidge 2011). This article has added to our understanding of participatory visual methods by exploring how cognitive mapping techniques can be used to understand people's attachment to a socializing place.

As this article illustrates, mental mapping can be a relatively simple, straightforward process. Over a twenty-year period of time, students at Kansas State University were asked to sketch their image and perceptions of Manhattan, Kansas. Of all the places represented, the vast majority of students' maps included the business district known locally as Aggieville (Table 1). "The Ville" is an organically created space where past and present students can purchase essential items, as well as gather to socialize. Our analysis of the mental maps reveals that students demonstrate all three types of attachment to place. Students are functionally attached because the business district located on the edge of campus fulfills nearly all their life-sustaining needs (e.g. food, clothing, housing, employment) enabling them to achieve their life goals. Students demonstrate territorial attachment because Aggieville is a space created by students, for students. The accepted norms and activities that dominate Aggieville (especially on Friday and Saturday nights) are intended for students only. Students have also established an emotional attachment to "The Ville" because it is an important place to socialize. Aggieville is a place to network and bond with peers, as well as a place to express their individual identity and even meet their life partner.

Our analysis of students' mental maps reveals that both individual experiences and collective group behavior are represented. Mental maps are unique to each individual, yet there is amazing consistency, repetition, and consensus among participants. When combined with the principles of psychogeography and walking interviews, an incredible amount of detailed information (e.g. backstory and context) about place is revealed. Whereas traditional ethnographic techniques (e.g. informal interviews) reveal the importance of particular places/locations, mental maps and walking interviews are effective in exposing gradients of emotional connection along the transect. For example, students' sketches showed that some parts of Aggieville are laden with intense emotional feelings (positive and negative), while other locations are emotionally neutral. We found that when a person is brought face-to-face with a particular location, intense emotions are triggered. Visiting the exact spot where a significant event unfolded elicits a deeper, richer flood of memories and emotions.

Lynch identified that landmarks are a key element of peoples' mental maps. In a more practical sense, mental mapping has the ability to show how peoples' interaction with a place changes over time. Over the course of two decades, it was fascinating to see that the core of Aggieville (Moro Street) has consistently been a popular "hang out" among students. Yet, the popularity of other places within Aggieville has ebbed and flowed over the decades. This is due in large part to changes in cultural norms and student preferences (e.g. disco versus country music), as well as high business turnover. A future project could map the evolving activity space and emotional connections students have over time.

Finally, there are a number of challenges associated with mental mapping and psychogeography, most of which are logistical in nature (see also Curtis 2016). This method of data collection tends to be demanding and time consuming for both the researcher and the participant. An effective, usable mental map can take a considerable amount of time for an individual to draw (as much as 30 to 40 minutes). It has proven helpful and efficient for a group of people to sketch their mental maps at the same time (e.g. focus group). Likewise, our walking interviews were conducted with a group of 12 individuals lasting over 90 minutes. We gathered a great deal of information, but our sample size was small. Had our goal been to collect a representative sample of the student body population, the amount of time needed to conduct such transects with a larger sample size would have been considerably greater. Furthermore, some individuals are geographically challenged and their mental maps are incomplete, at best. It can take considerable time and effort to help an individual understand what is expected of them as they sketch their perception of a place. Finally, longitudinal studies typically require considerable forethought and planning. Serendipitously, because the mental maps we utilized were generated as part of classroom assignments, we had twenty years' worth of mental maps at our disposal. Collecting maps from a wide variety of individuals afforded us a larger spread of cognitive impressions of place. Thus, although the maps don't always provide depth in individual observations, they did provide considerable breadth. Despite these challenges, mental mapping and walking interviews are extremely effective tools in assessing peoples' emotional connection to place.

Note

1. Jacqueline Curtis (2016) provides an excellent overview of the evolution of mental mapping.

Acknowledgments

The authors wish to thank the individuals who participated in the walking interview, Alex Diener and Joshua Hagen for coordinating the special issue on place attachment, and the anonymous reviewers for their insights and valuable contributions to this article.

References

ABA [Aggieville Business Association]. 2016. *State of the Ville: Aggieville Community Vision Plan Background Report*. City of Manhattan, Kans. https://cityofmhk.com/DocumentCenter/View/41541/State-of-the-Ville-FINAL?bidId=.

Barcus, H. R., and S. D. Brunn. 2009. Towards a Typology of Mobility and Place Attachment in Rural America. *Journal of Appalachian Studies* 15: 26–48.

Boschmann, E. E., and E. Cubbon. 2014. Sketch Maps and Qualitative GIS: Using Cartographies of Individual Spatial Narratives in Geographic Research. *Professional Geographer* 66 (2): 236–248.

Chamlee-Wright, E., and V. H. Storr. 2009. "There's No Place Like New Orleans": Sense of Place and Community Recovery in the Ninth Ward After Hurricane Katrina. *Journal of Urban Affairs* 31: 615–634.

Curtis, J. W. 2016. Transcribing from the Mind to the Map: Tracing the Evolution of a Concept. *Geographical Review* 106 (3): 338–359.

Economist. 2018. Special Report on Germany: Identity Whose Heimat? Economist, 14 April, 6.

Evans J., and P. Jones. 2011. The Walking Interview: Methodology, Mobility, and Place. *Applied Geography* 31 (2): 849–858.

Fauntleroy, G. 1997. La Querencia: The Place of Your Heart's Desire. *New Mexico Magazine* 75: 22–30.

Gumprecht, B. 2008. *The American College Town*. Amherst: University of Massachusetts Press.

Hanna, S. P., P. L. Carter, A. E. Potter, C. Forbes-Bright, D. A. Alderman, E. A. Modlin, and D. L. Butler. 2018. Following the Story: Narrative Mapping as a Mobile Method for Tracking and Interrogating Spatial Narratives. *Journal of Heritage Tourism* (April): 1–18. https://doi.org/10.1080/1743873X.2018.14459628.

Hernández, B., M. C. Hidalgo, and C. Ruiz. 2014. Theoretical and Methodological Aspects of Research on Place Attachment. In *Place Attachment: Advances in Theory, Methods, and Applications*, edited by L. Manzo and P. Devine-Wright, 125–137. London: Routledge.

Ingold, T., and J. Lee. 2008. *Ways of Walking: Ethnography and Practice on Foot*. London: Ashgate.

Jorgensen, B. S., and R. C. Stedman. 2006. A Comparative Analysis of Predictors of Sense of Place Dimensions: Attachment to, Dependence on, and Identification with Lakeshore Properties. *Journal of Environmental Management* 79: 316–327.
Kusenbach, M. 2003. Street Phenomenology: The Go-Along as Ethnographic Research Tool. *Ethnography* 4 (3): 455–485.
Lewicka, M. 2011. Place Attachment: How Far Have We Come in the Last 40 Years? *Journal of Environmental Psychology* 31: 207–230.
Livingston, M., N. Bailey, and A. Kearns. 2008. *People's Attachment to Place – The Influence of Neighborhood Deprivation*. York, U.K.: The Joseph Rowntree Foundation.
Lynch, K. 1960. *The Image of the City*. Cambridge, Mass.: The M.I.T. Press.
Manzo, L. C., and P. Devine-Wright. 2014. *Place Attachment: Advances in Theory, Methods, and Applications*. London: Routledge.
Massey, D. 2005. *For Space*. London: Sage
Milligan, M. J. 1998. Interactional Past and Potential: The Social Construction of Place Attachment. *Symbolic Interaction* 21: 1–33.
Mitchell, D. 2003. *The Right to the City: Social Justice and the Fight for Public Space in America*. New York: Guilford.
Morgan, P. 2010. Towards a Developmental Theory of Place Attachment. *Journal of Environmental Psychology* 30 (1): 11–22.
Oldenburg, R. 1989. *The Great Good Place: Cafes, Coffee Shops, Bookstores, Bars, Hair Salons, and Other Hangouts at the Heart of a Community*. New York: Marlowe and Company.
Patterson, M., and D. Williams. 2005. Maintaining Research Traditions on Place: Diversity of Thought and Scientific Progress. *Journal of Environmental Psychology* 25 (4): 361–380.
Relph, E. 1976. *Place and Placelessness*. London: Pion.
Rishbeth, C. 2014. Articulating Transnational Attachments through On-Site Narratives. In *Place Attachment: Advances in Theory, Methods, and Applications*, edited by L. Manzo and P. Devine-Wright, 100–111. London: Routledge.
Rowles, G. D. 1990. Place Attachment among the Small Town Elderly. *Journal of Rural Community Psychology* 11: 103–120.
Scannell, L., and R. Gifford. 2010. Defining Place Attachment: A Tripartite Organizing Framework. *Journal of Environmental Psychology* 30: 1–10.
Seamon, D. 2008. Place, Placelessness, Insideness, and Outsideness in John Sayles' Sunshine State. *Aether* 3: 1–19.
_____ 2014. Place Attachment and Phenomenology: The Synergistic Dynamism of Place. In *Place Attachment: Advances in Theory, Methods, and Applications*, edited by L. Manzo and P. Devine-Wright, 11–22. London: Routledge.
_____ 2018. *Life Takes Place: Phenomenology, Lifeworlds, and Place Making*. London: Routledge.
Self, W., and R. Steadman. 2007. *Psychogeography: Disentangling the Modern Conundrum of Psyche and Place*. New York: Bloomsbury.
Silver, D. A., and T. N. Clark. 2016. *Scenescapes: How Qualities of Place Shape Social Life*. Chicago, Ill.: University of Chicago Press.
Smith, J. S. 2018. *Explorations in Place Attachment*. New York: Routledge.
_____, and M. Cartlidge. 2011. Place Attachment among Retirees in Greensburg, Kansas. *Geographical Review* 101 (4): 536–555.
Tsai, B. W., and Y. C. Lo. 2013. The Spatial Knowledge of Indigenous People in Mountainous Environments: A Case Study of Three Taiwanese Indigenous Tribes. *Geographical Review* 103 (3): 390–408.
Tuan, Y.-F. 1974. *Topophilia: A Study of Environmental Perception, Attitudes, and Values*. Englewood Cliffs, N.J.: Prentice-Hall.
_____, 1980. Rootedness versus Sense of Place. *Landscape* 24: 3–8.
Walter, D. 1989. *Aggieville, 1889-1989: 100 Years of the Aggieville Tradition*. Manhattan, Kans.: Ag Press.

RURALITY AS A KEY FACTOR FOR PLACE ATTACHMENT IN THE GREAT PLAINS

ANDREW HUSA and CHERYL E. MORSE

ABSTRACT. Results from an online survey of the mobility of people who grew up in Nebraska indicate that rural stayers, defined as those Nebraskans who have had some experience living in communities with fewer than 2,500 residents, express stronger place-based attachments than others. Key components of this attachment include general contentment living in Nebraska, connection to family, a desire to raise children in Nebraska, appreciation for the state's culture and hospitality, and for some, a fondness for the state's natural environments. The research draws on quantitative evidence and the personal stories shared by survey respondents. The findings from the Nebraska Roots Migration Survey are provocative in their suggestion that those who grow up in America's rural areas may develop a different set of values and social and place-based attachments than their urban peers. *Keywords: rurality, stayers, place attachment, Great Plains, Nebraska.*

Nebraska occupies an empty space in the American geographical imagination. The entire Great Plains region may appear as "placeless" or even a "nonplace" to those who fly over or drive through it as quickly as possible to get from one coast to another (Cresswell 2015). Yet, research with people who have grown up and live in Nebraska now suggests that its rural communities are rich with social interaction and places to which residents are deeply attached. It is a place that many people have been happy to make their home. Drawing on results from an online survey of the mobility of people who grew up in Nebraska, this paper asserts that those who have chosen to remain in the state and who have experienced living in or around rural towns, defined as communities of under 2,500 people, are even more content to stay in Nebraska than those from urban communities. The findings from the Nebraska Roots Migration Survey are provocative in their suggestion that those who grow up in America's rural areas may develop a different set of values and social and place-based attachments than urban residents.

Literature Review

Immobility in Migration Research

Research on residential mobility has often focused on why people leave a place, rather than why they choose to stay. This is especially true regarding migration research on the Great Plains, where rural depopulation and the region have been seemingly synonymous with each other since the early twentieth century (Loveridge, Yi, and Bokemeier 2009; Gilbreath 2012; Jacquet, Guthrie, and

Jackson 2016). Subsequently, there is little literature on "stayers," the people who have chosen to stay in this rural landscape.

Stayers are defined as people who are born and raised in an area and have never left (Stockdale and Haartsen 2018). A stayer is someone who has made the decision to remain in a place. As many rural-migration researchers have shown, the decision to move or migrate is emotional, complex, and situational, resting on a diverse set of temporal, environmental, economic, personal, and social conditions (Ní Laoire 2000; Halfacree 2006; Corbett 2013). Until recently, less attention has been given to the conditions and decision-making processes that lead rural people to stay in their home places.

Staying has often been characterized as being "stuck," as an inability to move. Jorgen Carling's research (2002), for example, uncovered "involuntary immobility" among Cape Verdeans who wished to migrate but were unable to do so. Yet, new research with stayers has revealed other experiences of staying. For example, Lance Erickson and colleagues' (2018) study of stayers in Montana found that rather than being stuck, some rural stayers make a deliberate decision to stay where they are, and in so doing, contribute to small town resilience and quality of rural life. This work is one example of the ways in which migration researchers are beginning to consider immobility from multiple vantage points. In 2018, Aileen Stockdale and Tialda Haartsen called for the adoption of an immobility perspective in migration research. This perspective considers stayers as active participants in residential decisions and staying as an active process. Similarly, Cheryl Morse and Jill Mudgett (2018) have argued for the importance of examining immobility as a choice. Their research unpacked varied reasons why people choose to stay, highlighting the different forms of mobility that stayers exercise in their immobility. This paper aims to contribute to this growing body of work that elicits the nuanced and complex factors that contribute to immobility.

PLACE ATTACHMENT IN RURAL-MIGRATION RESEARCH

Nearly a half-century ago, Yi-Fu Tuan asserted that "Place is a center of meaning constructed by experience" (1975, 15). He argued that we know places, especially places like the home and neighborhood, through our experiential and sensory engagement with them, and such engagement produces meaning in people's daily lives. Tuan grappled with definitions of "rootedness" and "sense of place," suggesting that rootedness describes those who have remained in place and lack an ability to reflect on their place, whereas those of a certain mobility status can develop a broader sense of place (1980). In recent years, geographers have focused on the embodied dimension of place, and even mobile experiences of place, in a variety of settings and modes (see Cresswell 2015, for a review). Given the decades-long history of investigations into the meaning of place, it is surprising that the role of place-based experience and emotional attachment to particular geographies has not been a central concern in migration studies.

Place attachment has only recently entered into the mix of factors that migration researchers consider as they analyze factors for staying in rural places. This mix of factors includes the physical and social characteristics of place alongside the roles of family, friends, and community in developing a sense of home, belonging, and rootedness (Stockdale and Haartsen 2018; Stockdale, Theunissen, and Haartsen 2018). Family ties, especially, have been linked to place attachment to, or rootedness in, a community (Barcus and Brunn 2009; Clark, Duque-Calvache, and Palomares-Linares 2015). The broader network of social ties with communities of small population size may also enhance place attachment. In her examination of the strength of rural place attachments, Maria Lewicka (2005) found that residents of villages reported significantly higher place attachment, had stronger neighborhood ties, and expressed more interest in their roots compared to residents of bigger communities. Research on rural and urban dwellers suggests that people who live in the country are more attached to their homes and local areas than people in urban areas (Anton and Lawrence 2014; Mandal 2016). This supports Lewicka's (2005) finding that place attachment was related to community size and that one would expect people in rural communities to be more attached to their hometowns.

Holly Barcus' scholarship has shown that in places as culturally different as Appalachia and Mongolia, people who have grown up in rural areas and moved away retain their attachment to their home places (Barcus and Brunn 2009, 2010; Barcus and Werner 2017). Attachments to the rural community are an important source of legitimization and identity construction (Pedersen 2018). This suggests a rural childhood may construct a much different place-based identity and set of values about place than a nonrural childhood. Research with people who have out-migrated from Vermont showed that they retain a place-based rural identity even decades after leaving, and that this connection has endured for centuries (Morse and Mudgett 2017).

Similarly, research in rural communities of the Great Plains indicates that strong attachment to place informs the identities of those who have spent their lives in the region (Smith and McAlister 2015). Specifically, William Wetherholt (2016) found that rural Great Plains residents who live in small towns expressed strong community attachments, a sense of pride in the community, and a feeling of acceptance. New insights on the strong influence of place attachment on staying in Vermont prompted Cheryl Morse and Jill Mudgett (2018) to call for detailed examinations of place attachment across the urban-rural gradient. While there have been studies related to differences in values between rural and urban dwellers as middle-aged adults (Kasarda and Janowitz 1974; Erickson, Sanders, and Cope 2018) and senior citizens (Usha and Lalitha 2016; Anderson, Larkins, Beaney, and Ray 2018; Pereira, Fonseca, Lies, and Kang 2019), there has been little research into how growing up in a rural versus an urban environment has influenced different values.

The research at the center of this paper situates itself within new studies of differences in place attachment between rural and urban residents as it attempts to address questions about place attachment in the decision to stay in the rural Great Plains, and specifically, Nebraska.

Case Background

The state of Nebraska lies in the heart of the Great Plains. Although Nebraska is the 16th largest state by land area, it was only the 38th largest in population at the 2010 census. The state's physical landscape is divided between rolling hills and ranches in the west and flat (see Figure 1) and fertile land in the east. Several small communities dot the state's vast rural landscape (see Figure 2). While 94 percent of Nebraska's land is defined as rural or agricultural, only approximately 27 percent of the state's population call those areas home. The other 73 percent can be found in the state's urban communities, especially in its largest city, Omaha (pop. 408,958), and second largest and state capital, Lincoln (pop. 258,379) (see Figure 3). These are the only two cities in the state with populations that exceed 100,000 people. Outside of the state's ten largest communities, the rest have populations of under 20,000 people (Archer and others 2017). As of 2010, Nebraska had 48 communities that had populations of 2,500 or more people and 487 with populations of under 2,500 (Figure 4). It is

Fig. 1—Rural Landscape in Blaine County (Photo by Andrew Husa)

FIG. 2—Small Town Landscape in Wilber, pop. 1,855 (Photo by Andrew Husa)

FIG. 3—Urban Landscape in Lincoln, pop. 258,379 (Photo by Andrew Husa)

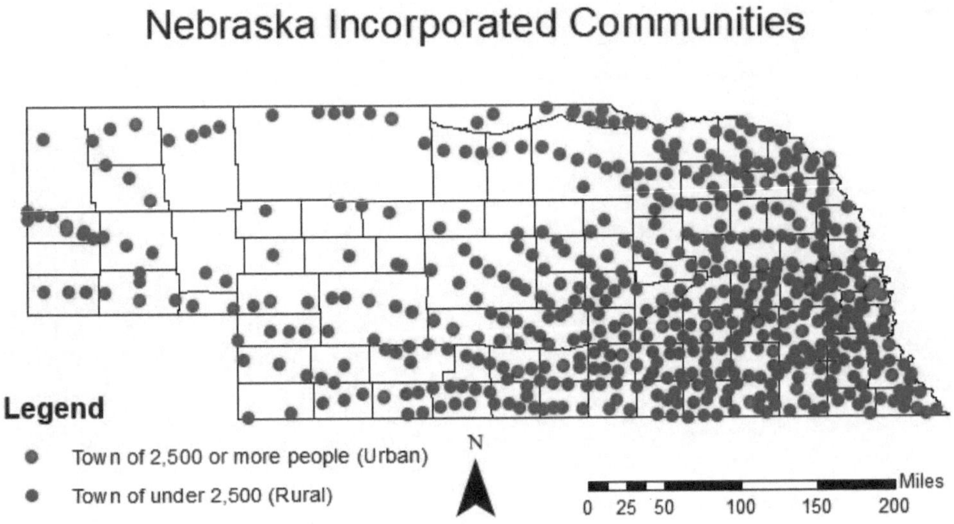

FIG. 4—Map of Nebraska Incorporated Communities (Map by Andrew Husa)

important to note that many rural residents live far from even the center of a small community, on ranches and farms distant from town. These residents are included in the rural census category.

METHODS

The Nebraska Roots Migration Survey builds upon an online residential decision-making survey that was first developed in 2014 (Morse 2015). The Vermont Roots Migration Survey was distributed on the social media networks of five individuals of differing ages who had attended a public high school while residing in Vermont. The aim of that survey, and the aim of the Nebraska Roots Migration Survey, was to ask people who had spent at least some portion of their adolescence in a specific state to explain in their own words why they chose to stay in, leave, or return to their home state. In addition to collecting narratives, the surveys gathered demographic data and offered a multiple-choice based list of factors for staying, leaving, or returning. For example, those people who identified as "stayers" were provided a list of reasons for staying that included choices such as "my work is in the state," "I would like to leave but I have not had the opportunity," "I appreciate the state's landscape (natural environments)," and "I have stayed in the state to live near family." A write-in box for "other" was offered. Respondents were free to select as many factors that fit their experience as they liked.

The unexpectedly high response rate to the 2014 Vermont survey—3,692 fully complete responses in a three week period (see Morse 2015; Morse and Mudgett 2018)—indicated that using social media networks of a targeted population was an effective method for efficiently generating a large data set that yielded both qualitative and quantitative data about mobility and immobility decisions. Although the Vermont survey was not intended to, and did not ultimately capture, a representative sample of the Vermont population, the high number of responses allowed for between-group comparisons by gender, educational attainment, and other areas of difference.

The Nebraska Roots Migration Survey used the original set of questions in the Vermont survey and included several additional questions, such as income level, religious affiliation, and military service. While the Vermont survey asked respondents to name the town they resided in while attending high school, the Nebraska survey went one step further by asking respondents to describe their hometowns in their own words. Hometown characteristics became a focus of the analysis of the resulting Nebraska data (Husa 2020), although they were not the focus of this paper.

The Nebraska Roots Migration Survey questions were edited from their original Vermont versions, formatted, and then uploaded onto Lime Survey (cloud-based software) to distribute the survey. The research was reviewed and received approval from the University of Nebraska-Lincoln's institutional review board. The first two questions of the survey asked whether the respondent had attended some high school in Nebraska and if they were 19 years of age and older. It the answer to either was "no," the respondent was exited from the survey. Four people who had grown up in Nebraska, two men and two women, ranging in age from late twenties to early forties, and who had attended high schools in different regions of the state posted the web link to the survey on their social media sites, primarily Facebook pages. The goal was to tap into the social networks of people of different age groups and geographic origins within the state. They invited others to pass on the link to others who had attended at least some high school in Nebraska. The survey was anonymous, but those who wished to be contacted for further research were invited to give their contact information at the end of the survey. The survey ran for three weeks in February and March 2019.

The Nebraska Roots Migration Survey gathered 1,827 fully completed responses. Each of the state's 93 counties was represented in the data set. The survey responses do not constitute a representative sample of Nebraska's population. A higher percentage of women responded to the survey (61.2 percent) than the percentage of women in Nebraska's overall population (50.2 percent, US Census 2013–2017 American Community Survey). A higher percentage of survey respondents self-identified as "white" (93.1 percent) compared to the state estimate of persons who self-identified as "white" (87.8 percent). While

24.7 percent of the survey respondents were aged 25–34, this age group only made up 13.4 percent of the state's population in the 2013–2017 American Community Survey. Percentages among income brackets were similar between survey respondents and the state estimate. Education percentages were much different, with the percentage of survey respondents who obtained their bachelor's degree nearly 15 percent higher than the percentage reported by the American Community Survey (2013–2017). As the survey did not aim to be representative of the state's population and because we conducted comparative analyses between groups, the influence of these imbalances on our findings is minimal. As we discuss in the findings, stayer survey respondents expressed favorable attitudes towards living in Nebraska. As respondents were encouraged to pass the survey link on to others, it is possible that respondents extended the survey to individuals who might give favorable responses. Or, it could be that people who were more content with their residential decisions chose to take the survey. These are some of the limitations that are inherent to open access surveys.

The quantitative data were analyzed using the SPSS (version 25) statistical analysis program (IBM Corp. SPSS 2017). Between-group comparisons of migration factors were made using a cross-tabulation analysis; chi-square test results with a p-value of less than or equal to 0.05 were considered significant. Qualitative data, which included respondents' narratives, were coded thematically using the Dedoose software package. The text was first analyzed for themes developed from the statistical analysis. In the second round of coding, nuances related to the original themes and emergent codes were identified. The analysis therefore was driven by the findings from the quantitative analysis and the analysis of narratives was conducted to bring a "biographical approach" to understandings of how complex residential and mobility decisions are made by individuals within the context of their particular circumstances (Halfacree and Boyle 1993; Halfacree and Merriman 2015; Morse and Mudgett 2018).

Definitions

The United States Census Bureau does not specifically define "rural"; instead a rural place is defined as an absence of the urban (United States Census Bureau 2020). "Rural" encompasses all population, housing, and territory not included in an urban area that is defined as a having at least 2,500 people (United States Census Bureau 2020). Census categories do not always adequately describe the lived experience of a place. As Michael Woods (2011) has explained, rural is an elusive concept; around the world, demographers, geographers, and government leaders struggle to develop definitions of rurality that are both uniform and conform to personal experience of living in a place. Rural is relative. For the person living on a farm many miles from a town of 10,000, that town may feel urban, and to the urbanite, a trip to the suburbs may feel like a trip to the countryside. How rurality is experienced as an identity, a place, and a representation also varies person to person and culturally (Halfacree 2006). For the purposes of clarity, we adopt the U.S. Census definition of rural in this analysis.

In our analysis, survey respondents from towns of 2,500 or more people are classified as urban while those from towns of under 2,500 are classified as rural. As of 2010, Nebraska's population of 1,826,341 was comprised of 490,655 (26.9 percent) rural and 1,335,686 (73.1 percent) urban inhabitants. The state's population has experienced slow growth in the last decade. The U.S. Census' 2013–2017 American Community Survey estimated Nebraska's population as 1,893,921, with most of that growth taking place in metropolitan counties (Olberding 2019).

For the purposes of this paper, we applied the U.S. Census' definitions to code all communities with fewer than 2,500 residents as rural and those with 2,500 or more residents as urban to allow for between-group comparisons of those who grew up and/or presently live in (or near) rural communities with respondents who did not or do not. We applied the 2013–2017 American Community Survey's population estimates in order to use the most current data. Those survey respondents who confirmed that they had never lived outside of Nebraska, apart from schooling or military service, for a full year or more are referred to as "stayers." "Leavers" are survey respondents who have moved outside of the state and have not returned to reside in Nebraska. Those who have lived outside of Nebraska for at least one year and then returned to live in Nebraska are referred to as "returnees."

This paper focuses on experiences and place attachments of the Nebraska stayers who responded to the survey. Keen to consider the ways in which rural residential experience might relate to reasons for staying in Nebraska, and to place attachments specifically, we developed three different means to track possible rural/urban difference. First, we coded the hometown of each respondent as rural or urban. This allowed us to discern any effects that growing up in a small community may have on residential choices. We refer to participants as "rural hometowners" and "urban hometowners." We also coded the present community in which each stayer lives as urban and rural, allowing for a rural/urban comparison of the settings in which participants have chosen to reside. These groups are called "rural residents" and "urban residents." Finally, we identified those people who both grew up in a rural hometown and live now in a rural town (not necessarily the same community). We refer to these people as "all-rural stayers." Respondents who have never lived in a rural town, those who lived in a rural hometown but now live in an urban place, and those who grew up in an urban town or city but now live in a rural town are referred to as "not all-rural stayers."

TABLE 1—REASONS FOR CITED FOR STAYING IN NEBRASKA AS A PERCENT OF ALL STAYERS

REASONS FOR STAYING IN NEBRASKA	PERCENT (%) OF ALL STAYERS WHO SELECTED REASON
I stayed in NE to live near family	73.9
I enjoy living in Nebraska (NE)	73.0
I appreciate NE's culture/community	54.2
I like NE's rural hospitality	53.2
I wanted to raise my children in NE	47.9
I like NE's landscape (natural environments)	44.3
My work is based in NE	42.6
I never wanted to live outside NE	33.5
I have visited other places and decided NE is the best place for me	18.6
Partner didn't want to leave NE	15.1
I would like to leave NE but have not had the opportunity	15.4
I stayed in NE to care for family members	11.9
NE offers me activities and opportunities that don't exist elsewhere	8.6

FINDINGS

The Nebraska Roots Migration Survey received a total of 1,827 fully completed responses. Of these, 56.0 percent (1,024) were from stayers. About half of the stayer respondents (56 percent, n=517) grew up in a rural hometown of fewer than 2,500; a third of respondents (33.3 percent, n=340) lived in a rural town at the time of the survey. Two-hundred and sixty-five respondents (25.9 percent of all stayers) are all-rural stayers who both grew up in and now live in a community of fewer than 2,500 people. Table 1 shows the overall results of reasons for staying in Nebraska. Similar to the Vermont Roots Migration Survey results, hometown stayers showed a high level of agreement about why they stayed and expressed contentment about living in their home state (Morse and Mudgett 2018).

The large majority of Nebraskan stayers reported that they enjoy living in the state (73 percent) and that a desire to live near their family was a key reason for staying. Other reasons for staying that were true for more than half of the stayers were an appreciation for the state's culture and community (54.2 percent) and appreciation for Nebraska's rural hospitality (53.2 percent). Nearly half (47.9 percent) of stayers said they wanted to raise their children in Nebraska. Having work or a job in Nebraska was a reason for staying for 42.6 percent of respondents, and an appreciation for Nebraska's landscape or natural environment was cited as true by 44.3 percent of respondents. In the analysis section of this paper, we flesh out these findings thematically, adding in the diverse, emotionally rich, nuanced, and often place-based narratives about staying that Nebraskan stayers shared with us.

TABLE 2—REASONS FOR STAYING IN NEBRASKA BY RURAL LIVING EXPERIENCE

REASONS FOR STAYING IN NEBRASKA	PERCENT OF RESPONDENTS BY RURAL LIVING EXPERIENCE									
	ALL STAYERS N=1024	RURAL HOME-TOWNERS N=507	URBAN HOME-TOWNERS N=517	CHI-SQ P VALUE	RURAL RESIDENTS N=340	URBAN RESIDENTS N=680	CHI-SQ P VALUE	ALL-RURAL STAYERS N=265	NOT ALL-RURAL STAYERS N=757	CHI-SQ P VALUE
I stayed in NE to live near family	73.9	76.2	71.6		73.8	74.3		74.7	73.7	
I enjoy living in Nebraska (NE)	73.0	77.9	68.0	.000	77.9	70.6	.013	79.6	70.7	.005
I appreciate NE's culture/community	54.2	60.9	47.3	.000	61.5	50.6	.001	64.5	50.5	.000
I like NE's rural hospitality	53.2	65.6	40.6	.000	70.6	44.4	.000	72.5	46.4	.000
I wanted to raise my children in NE	47.9	56.1	39.4	.000	60.3	41.6	.000	62.3	42.8	.000
I like NE's landscape (natural environments)	44.3	50.5	38.1	.000	49.4	41.6	.018	51.3	41.7	.007
My work is based in NE	42.6	43.1	42.0		37.1	45.4	.011	38.1	44.1	
I never wanted to live outside NE	33.5	42.4	24.5	.000	42.1	29.3	.000	44.9	29.5	.000
I have visited other places and decided NE is the best place for me	18.6	19.0	18.1		21.8	17.1		22.3	17.3	
I would like to leave NE but have not had the opportunity	15.4	9.5	21.5	.000	7.4	19.3	.000	6.4	18.6	.000
Partner didn't want to leave NE	15.1	17.0	13.2		18.5	13.5	.036	18.1	14.1	
I stayed in NE to care for family members	11.9	11.4	12.4		10.3	12.8		10.2	12.5	
NE offers me activities and opportunities that don't exist elsewhere	8.6	8.7	8.5		9.7	8.1		10.6	7.9	

Rurality appears to be an important factor in both staying and place attachment among survey respondents. Respondents who grew up in a rural hometown were more likely than urban hometowners to stay in Nebraska (61.9 percent as compared to 51.1 percent). Digging deeper into the quantitative analysis of reasons for staying, we conducted cross-tabulation comparisons of rural to urban hometowners, rural to urban residents, and all-rural to not all-rural stayers (Table 2). We found strong correlations between rurality and specific values and social and environmental perceptions. To generalize, it appears that respondents who have lived in small towns have stronger place-based attachments to Nebraska than those who have not. The desire to raise children in Nebraska as a factor for staying offers one example. Of all stayers, 47.9 percent indicated that a desire to raise their children in Nebraska was a reason for them to stay. However, those who grew up in rural hometown were much more likely to list this as a factor than urban hometowners (56.1 percent vs 39.4 percent, p=.000). A similar difference was found among those who now live in small towns, who were more likely than those who live in urban areas to select this reason (60.3 percent vs 41.6 percent, p=.000). This pattern repeats among all-rural stayers, who were the most likely to say they wished to raise their children in Nebraska, compared to the not all-rural stayers (62.3 percent vs 42.8 percent, p=.000).

To summarize Table 2, in our group of Nebraska stayers, people with some rural living experience were more likely to express contentment living in Nebraska, and appreciation for its culture and community, rural hospitality, and natural environments. They were more likely to say that they wanted to raise their children in the state. They are less likely than those without rural living experience to say they would like to leave Nebraska. In the following section, we turn to respondents' stories to gain deeper insight into rural place-based attachments.

The results of the Nebraska Roots Migration Survey show several connections between the residents and their hometowns, and that rural Nebraska is a place that people happily choose to stay. At the same time, the survey results show the difference in values between those have lived in rural areas and those who have not, as well as the different considerations that must be made in both rural and urban settings in residential decision making. We pursued some of the findings thematically by analyzing how stayers articulated reasons for staying in their own words.

Familial Attachment

Staying to live near family was, by far, the most agreed upon reason for staying in Nebraska. Family appears to be a key factor for binding people to place, especially for those who grew up and live now in a rural town. This supports research that lists family ties among the most important measures of place attachment (Barcus and Brunn 2009; Clark, Duque-Calvache, and Palomares-

Linares 2015). A large majority of stayers, 98.4 percent, have family members who live in Nebraska. The emotional attachment to family, as well as to the connection between ancestors and Nebraska came through in the stories survey respondents shared.

When asked to describe in their own words why they chose to stay in Nebraska, respondents frequently referred to their family ties to the state. In many cases, these rural respondents stated that they were fifth- or sixth-generation Nebraskans, and that they grew up in the same town as their ancestors and still hold tight to their ancestral roots. These accounts referenced Euro-American settler history and land acquisition.[1]For example, a woman from Alexandria expressed this heritage-oriented familial attachment: "My ancestors came to Nebraska to live before Nebraska was even a state. We have deep roots here." Similarly, a Sargent resident said, "I'm a 5th generation Nebraska farmer and so is my husband. Nebraska is our roots. I can't imagine living anywhere else." This response from a woman who lives in Julian indicates the strong tie between place and family: "My ancestors on my dad's side homesteaded here so our roots are here and I take pride in that. I don't plan to ever leave Nebraska because my roots are deep here." In each of these responses, there is a sense that rootedness to land is a dimension of family heritage and identity.

For other respondents, the idea of family expresses a way of living that is not necessarily tied to land. Perhaps one man from the small town of Filley summed it up best with his written response: "Family is one of my most deep-seeded values and to leave the state would mean to leave almost all of my family, and thus, those values. I couldn't picture not being in Nebraska, where I am close to my family."

Stayers with rural living experience were statistically more likely to consider child raising in their decision to stay in Nebraska. Many written responses from those who said they wanted to raise their children in Nebraska made mention of their family ties in this decision. This decision is based on both heritage and community pride, with respondents writing that they want to raise their children near family and continue the heritage their parents, grandparents, and even great grandparents created in the community. Several noted that they wanted to raise their children in the same community that they were raised in, as it was important to many rural hometown stayers that their children have the same positive childhood experiences that they had. Including both familial attachment and attachment to Nebraska's community and culture in her response, a young woman from Tobias wrote that she "Decided to stay close to my roots and family to raise my children to work hard and be a part of a community. A place to teach future generations about love, respect, and hard work."

Attachment to Rural Culture and Hospitality

A majority of Nebraska stayer respondents expressed appreciation for the culture and social networks present in Nebraskan communities. Several respondents placed a high value on working hard, community pride, familiarity with neighbors, and of course living next to friends and family in their decisions to stay in Nebraska. This local attachment is the result of relationships that have developed over time, as family ties and general familiarity have led to many rural residents having a strong sense of belonging to the community. The results of the Nebraska Roots Migration Survey support Monica Lewicka's (2005) research on the relation of place attachment and community size, and reaffirm her findings that people in rural communities are more attached to their hometowns.

Nebraska's rural hospitality and culture and community were cited by several rural respondents. Several of the written responses included references to the University of Nebraska fight song: "There is no place like Nebraska. The people, the community, the lifestyle and values are wonderful" and "Not only have I chosen to live in Nebraska, but also in the same community. I have traveled to Europe twice and in most of the states, but return here as like the song says: 'there is no place like Nebraska.'"

Attachment to Landscape/Environment

Nebraska survey respondents reported fewer instances of physical place attachment when compared to research on the rural population in other states, such as Vermont (Morse and Mudgett 2017) or Montana (Erickson, Sanders, and Cope 2018). However, appreciation for the landscape was cited more frequently by those with rural residential experience (Table 2). Just over one half (51.3 percent) of the all-rural stayers listed the landscape as a reason for staying in Nebraska, compared to 41.7 percent of the not all-rural stayers.

Several respondents specifically mentioned appreciation for the Nebraska Sandhills, an area of sprawling sand dunes that cover much of north-central Nebraska. The ties to the physical landscape appear to be primarily aesthetic, with respondents discussing the positive emotions and thoughts evoked by the scenery, as in these responses: "There is so much beauty around us every day; the native grasses in the road ditches, rolling hills planted with corn and beans, the Sandhills" and "I live on the edge of the Sandhills and, to me, it is one of the most beautiful places on earth. Love the rural pastures, cattle, and amazing skies!"

The openness of the sky was a feature admired by several respondents. This is a place-based value that is geographically specific to regions with broad vistas, such as the Great Plains. One woman from Verdigre described the state's landscape and environment, and the people who call it home, by saying, "Nebraska has things you can't put a price tag on. Beautiful sunrises and sunsets, the landscapes, the sounds of nature, the great people of this state, the opportunities for young and old, peace."

General Contentment

Some of the most significant differences between rural and urban residents are evident in the results of how one feels about living in Nebraska. Stayers with rural living experience were much more likely to enjoy living in the state, as well as much less likely to want to live permanently outside of Nebraska. Accordingly, many of the written responses related to feelings of general contentment from living in Nebraska were from rural stayers. For example, a respondent from Oakland wrote:

> It is my home. I have visited 49 states (all except Alaska) and 9 countries, staying in some places for up to 3 months but nothing compares to Nebraska. I love the seasons, the smells, the hospitality, football Saturdays, how the farmers have coffee together, and that I can drive to see most of my family within hours. I can have an uplifting conversation with a complete stranger at a grocery store. I have neighbors who will rake my leaves or shovel my driveway while I'm at work. I have access to all the amenities that a big city has to offer but it's mixed with a small town vibe. I love that I can look out my parent's picture window and see miles of fields and rolling hills. And the sunsets are to die for. There's no place like Nebraska.

Discussion

Analysis of Nebraskans' reasons for staying revealed that staying is an active decision involving a mix of complex factors including family ties of different kinds over varying lengths of time, cultural characteristics, social networks, work, and emotional attachment to place. Respondents' narratives weave together multiple conditions and reasons to explain immobility. In contrast to Tuan's (1980) conceptual distinction between rootedness and sense of place, the Nebraska survey respondents clearly articulated reflection on their attachments, and the reasons for their attachments to their home place. Many explained their rootedness as a way to show how they arrived at a sense of place. "Roots" was a word commonly used by stayers to describe their attachment. We are convinced that, rather than indicating an unreflexive connection to community and landscape, rootedness describes an emotional attachment born out of time spent in place and reflection on that experience.

Family and community engagement emerged as dominant factors for staying as indicated by the Nebraska Roots Migration Survey. The majority of Nebraskan stayers indicated that family ties were very important to their decision to stay. Therefore, our research aligns with other rural migration scholarship that has demonstrated family is an important factor in residential decision making (Barcus and Brunn 2009; Clark, Duque-Calvache, and Palomares-Linares 2015; Morse and Mudgett 2018). However, our results also provide evidence that ties to community and social interactions, and a particular way of behaving in the world—hospitality—are highly valued by stayers. These topics require further research, as connection to community as apart from family may reveal the ways in which both mobile and immobile populations develop attachment to place. Such work may also go a long way in challenging narratives of rural places as empty or placeless, where nothing at all happens. The stories generated in the Nebraska Roots Migration Survey suggest that small communities are the sites for rich sets of social interactions that, aligning with Tuan's (1975) conception of place, produce meaning. Further, the research showed that the production of attachment to place may not be evenly distributed across geographical settings. We focus on this particular finding in the remainder of our discussion.

The Nebraska Roots Migration Survey produced compelling insight into the role that rural living may have on the development of strong attachment to place. It suggests that people who lived in a rural place during their adolescence and those who choose to live in a small town as adults are more likely to enjoy living in, value raising their children in, and appreciate the culture, community, and landscape of their home state more than those who do not have rural living experience. In short, the people with rural living experience in Nebraska were more content in their residential location than others. This research indicates that a childhood spent in a rural place impacts contentment in staying, even if the individual chooses to live in an urban setting in Nebraska as an adult. It shows that people who have lived in rural communities find it very important for their children to grow up in Nebraska's countryside places. These findings open up new areas for research and invites the generation of hypotheses into the relationships between rurality and place attachment. Following on Lewicka (2005), what is the role of scale of human settlement in attachment to place? How do rural childhoods differ from urban childhoods in terms of experience of place and mobility decisions in adulthood? How might small hometowns with different characteristics influence these findings? Is there a relationship between the time spent in a rural environment or community and the development of a strong attachment to place?

Tuan wrote, "Experience takes time. Sense of place is rarely acquired in passing. To know a place well requires long residence and deep involvement" (1975, p. 164). Our findings gesture toward a possible correlation between those who have lived the majority of their lives in one state to contentment, but is attachment to place an outcome of time spent in place, or a thoughtful reflection of engagement with place? Or, does rural living offer a qualitatively different kind of deep involvement with place, one that engages humans with environmental features and nonhuman actors, and one that builds an entirely different sense of place?

Most Americans live in urban areas; therefore, most have not had access to the kinds of experiences and conditions that the rural Nebraskan research participants so highly valued. As the Great Plains and other rural areas of the United States continue to hollow out, what is lost in terms of contented rural immobility?

Conclusions

Moving forward, we recommend further study in three areas. We call first for more research on the rural-urban dimensions of place-based attachment. Second, we encourage rural mobility and immobility researchers to continue their investigations into many facets of migration and emotion: returnees' experience of moving home, the values and feelings of urban hometowners who become rural residents, and the emotional geographies of rural living along many axes of difference including gender, age, race, sexual orientation, class, and residential history. Finally, we see a need for coordinated research projects that allow for in-depth qualitative research with rural residents along an extended set of research sites, creating a means to compare differences among place. We suggest that the outcome of such research will be of interest not only to the scholarly community, but to those who live and work in rural communities. Rural migration research will benefit individuals and groups who are concerned about the vitality of rural communities, as mobility research sheds light on the reasons why people choose to stay in or leave small towns. Our study supports other research that demonstrates strong place attachment among residents of rural communities (Lewicka 2005; Anton and Lawrence 2014; Smith and McAlister 2015; Mandal 2016). These residential preference data, which include both the positives and negatives of living in a rural community, can be applied to efforts to not only sustain the current population of rural communities, but to attract new residents. This information is consequential for rural geographers and migration experts who study a region such as the Great Plains, where rural depopulation has endured, but deep-rooted attachment to place remains.

By adopting an immobility perspective (Stockdale and Haartsen 2018) and examining the immobility of Nebraska's stayers as a choice (Morse and Mudgett 2018), our analysis revealed the varied reasons why people choose to stay in their home state. These people who grew up in and choose to live in a rural place were much more likely to enjoy living in Nebraska. The Nebraska Roots Migration Survey found that familial attachment and the state's rural culture and hospitality were the key drivers in rural Nebraskans' decisions to stay in their hometowns, or in another rural community. For these rural residents, the imagined emptiness of Nebraska's landscape is full of rich social interactions.

Note

[1] A total of three Nebraskan stayers identified as Native American. Two of them named family as reasons for staying, but neither mentioned generational or ancestral ties.

Orcid

Cheryl E. Morse http://orcid.org/0000-0001-9034-6459

References

Anderson, E., S. Larkins, S. Beaney, and R. Ray. 2018. Should I Stay or Go: Rural Ageing, a Time for Reflection. *Geriatrics* 3: 49. doi: 10.3390/geriatrics3030049.

Anton, C. E., and C. Lawrence. 2014. Home Is Where the Heart Is: The Effect of Residence on Place Attachment and Community Participation. *Journal of Environmental Psychology* 40: 451–461. doi: 10.1016/j.jenvp.2014.10.007.

Archer, J. C., and others. 2017. *Atlas of Nebraska*. Lincoln: University of Nebraska Press.

Barcus, H., and C. Werner. 2017. Choosing to Stay: (Im)mobility Decisions amongst Mongolia's Ethnic Kazakhs. *Globalizations* 14 (1): 32–50. doi: 10.1002/psp.2148.

Barcus, H. R., and S. D. Brunn. 2010. Place Elasticity: Exploring a New Conceptualization of Mobility and Place Attachment in Rural America. *Geografiska Annaler Series B-Human Geography* 92B (4): 281–295. doi: 10.1111/j.1468-0467.2010.00353.x.

Barcus, H., and S. D. Brunn. 2009. Towards a Typology of Mobility and Place Attachment in Rural America. *Journal of Appalachian Studies* 15 (1/2): 26–48.

Carling, J. 2002. Migration in the Age of Involuntary Immobility: Theoretical Reflections and Cape Verdean Experiences. *Journal of Ethnic and Migration Studies* 28 (1): 5–42. doi: 10.1080/13691830120103912.

Clark, W. A. V., R. Duque-Calvache, and I. Palomares-Linares. 2015. Place Attachment and the Decision to Stay in the Neighbourhood. *Population, Space and Place* 23 (2). doi: 10.1002/psp.2001.

Corbett, M. J. 2013. I'm Going to Make Sure I'm Ready Before I Leave: The Complexity of Educational and Mobility Decision-Making in a Canadian Coastal Community. *Journal of Rural Studies* 32: 275–282.

Cresswell, T. 2015. *Place: An Introduction*, 2nd ed. Malden, Mass.: John Wiley & Sons Ltd.

Erickson, L. D., S. R. Sanders, and M. R. Cope. 2018. Lifetime Stayers in Urban, Rural, and Highly Rural Communities in Montana. *Population, Space and Place* 24 (4). doi: 10.1002/psp.2133.

Gilbreath, A. 2012. "A Little Place Getting Smaller": Perceptions of Place and the Depopulation of Gove County, Kansas. *Great Plains Quarterly* 32 (1): 25–41.

Halfacree, K. 2006. Rural Space: Constructing a Three-fold Architecture. In *Handbook of Rural Studies*, edited by P. J. Cloke, T. Marsden, and P. H. Mooney, 44–62. London and Thousand Oaks, Calif.: Sage.

_____., and P. J. Boyle. 1993. The Challenge Facing Migration Research: The Case for a Biographical Approach. *Progress in Human Geography* 17 (3): 333–348.

_____., and P. Merriman. 2015. Performing Internal Migration. In *Internal Migration: Geographical Perspectives and Processes*, edited by D. P. Smith, N. Finney, K. Halfacree, and N. Walford, 149–164. Surrey, U.K., and Burlington, Vt.: Ashgate. doi: 10.4324/9781315589275.

Husa, A. 2020. Population Sustainability in Rural Nebraska Towns. Ph.D. diss., University of Nebraska-Lincoln.

IBM Corp. Released 2017. IBM SPSS Statistics for Windows, Version 25.0. Armonk, NY: IBM Corp.

Jacquet, J. B., E. Guthrie, and H. Jackson. 2016. Swept Out: Measuring Rurality and Migration Intentions on the Upper Great Plains. *Rural Sociology* 82 (4): 601–627. doi: 10.1111/ruso.12145.

Kasarda, J. D., and M. Janowitz. 1974. Community Attachment in Mass Society. *American Sociological Review* 39 (3): 328. doi: 10.2307/2094293.

Lewicka, M. 2005. Ways to Make People Active: The Role of Place Attachment, Cultural Capital, and Neighborhood Ties. *Journal of Environmental Psychology* 25 (4): 381–395. doi: 10.1016/j.jenvp.2005.10.004.

Loveridge, S., D. Yi, and J. L. Bokemeier. 2009. Why Are They Moving Away? Comparing Attachment to Place in the Great Plains to the Rest of the Nation. *Online Journal of Rural Research & Policy* 4 (1). doi: 10.4148/ojrrp.v4i1.56.

Mandal, A. 2016. Size and Type of Places, Geographical Region, Satisfaction with Life, Age, Sex and Place Attachment. *Polish Psychological Bulletin* 47 (1). 1: 159–69. doi: 10.1515/ppb-2016-0018.

Morse, C. E. 2017. The Emotional Geographies of Global Return Migration to Vermont. *Emotion, Space and Society* 25: 14–21. doi: 10.1016/j.emospa.2017.09.007.

_____. 2015. The Risks and Rewards of Using Social Media in Rural Migration Research: Findings from the Vermont Roots Project. *The Northeastern Geographer* 7: 72–88.

Morse, C. E., and J. Mudgett. 2018. Happy to Be Home: Place Attachment, Family Ties, and Mobility of Contented Rural Stayers. *The Professional Geographer* 0 (2): 261–269. doi: 10.1080/00330124.2017.1365309.

_____, and _____. 2017. Longing for Landscape: Homesickness and Place Attachment Among Rural Out-Migrants in the 19th and 21st Centuries. *Journal of Rural Studies* 50: 95–103. doi: 10.1016/j.jrurstud.2017.01.002.

Ní Laoire, C. 2000. Conceptualising Irish Rural Youth Migration: A Biographical Approach. *International Journal of Population Geography* 6 (3): 229–243. doi:10.1002/1099-1220(200005/06)6:3<229::AID-IJPG185>3.0.CO;2-R

Olberding, M. 2019. "Census: Nebraska's Big Counties Keep Growing, While Rural Counties Decline." *Lincoln Journal Star*. 18 April.

Pedersen, H. D. 2018. Is Out of Sight out of Mind? Place Attachment among Rural Youth Out-Migrants. *Sociologia Ruralis* 58 (3): 684–704. doi: 10.1111/soru.12214.

Pereira, G. F, M. Lies, and M. Kang. 2019. A Case Study of Place Attachment in Rural and Urban Senior Cohousing Communities. *Housing and Society* 46 (1): 3–22. doi: 10.1080/08882746.2019.1580945.

Smith, J. S., and J. M. McAlister. 2015. Understanding Place Attachment to the County in the American Great Plains. *Geographical Review* 105 (2): 178–98. doi: 10.1111/j.1931-0846.2014.12066.x.

Stockdale, A., and T. Haartsen. 2018. Editorial Introduction: Putting Rural Stayers in the Spotlight. *Population, Space and Place* 24 (4). doi: 10.1002/psp.2124.

Stockdale, A., N. Theunissen, and T. Haartsen. 2018. Staying in a State of Flux: A Life Course Perspective on the Diverse Staying Processes of Rural Young Adults. *Population, Space and Place* 24 (8). doi: 10.1002/psp.2139.

Tuan, Y. F. 1975. Place: An Experiential Perspective. *Geographical Review* 65 (2): 151–165. doi: 10.2307/213970.

_____ 1980. Rootedness Versus Sense of Place. *Landscape* 24 (1): 3–8.

United States Census Bureau; American Community Survey, 2017 5-Year Data Profile, DP05.

United States Census Bureau. "Urban and Rural." https://www.census.gov/programs-surveys/geography/guidance/geo-areas/urban-rural.html

Usha, V., and K. Lalitha. 2016. Quality of Life of Senior Citizens: A Rural-Urban Comparison. *Indian Journal of Social Psychiatry* 32 (2): 158. doi: 10.4103/0971-9962.181104.

Wetherholt, W. A. 2016. Exploring Rootedness in the Very Rural Great Plains Counties of Kansas and Nebraska. Ph.D. diss., Kansas State University.

Woods, M. 2011. *Rural*. New York and London: Routledge.

WHOSE PUGET SOUND?: EXAMINING PLACE ATTACHMENT, RESIDENCY, AND STEWARDSHIP IN THE PUGET SOUND REGION

DAVID J. TRIMBACH, WHITNEY FLEMING and KELLY BIEDENWEG

ABSTRACT. Place attachment is recognized as integral to addressing place-based challenges, such as environmental degradation. In combination with place identity and place meaning, place attachment can contribute to one's engagement in environmental stewardship behaviors. Building upon ongoing efforts to monitor the health and recovery of Washington's Puget Sound, a region experiencing profound population growth, the authors examine the relationships among place attachment, residency, and environmental stewardship. Based on data from over 2,000 responses to a general population survey, the authors highlight the current status of place attachment among Puget Sound residents and the extent to which residency matters to their stated attachment to place and environmental stewardship behaviors. This examination challenges often touted negative perceptions of the region's newcomers and concludes that residents, new and old, share a strong positive place attachment and sense of proenvironmental stewardship. *Keywords: place attachment, stewardship, residency, ecosystem recovery, Puget Sound.*

Washington's Puget Sound region, anchored by Seattle, has experienced remarkable growth over the last decade. Long associated with economic booms and busts, including those associated with natural resource industries, commercial aircraft production, and dot-coms, Puget Sound is a growing center for technological innovation (Morrill and Sommers 2005; Florida 2019). Major technology companies, like Amazon and Microsoft, have agglomerated to call this area home. Amazon alone occupied 19 percent of Seattle's prime office space as of 2017 and employed approximately 45,000 people out of Seattle's estimated 700,000 population (Holder 2019). Many of Amazon's employees, like other expanding companies, are new residents, extracted from elsewhere or attracted to the region because of job potential, recreation, and the environment, among other factors (Atlantic Media/Allstate 2017). Seattle is consistently ranked as one of the fastest growing cities in the country because of this growth (Brownstein 2017). While growth is concentrated in Seattle, according to Washington's Office of Financial Management (OFM), Puget Sound's twelve counties have all experienced population growth over the last decade (OFM 2017).

While desirable among governments and industries, the expansion of this "digital city" (Morrill and Sommers 2005) has negative impacts. For example, Puget Sound, particularly Seattle, is facing an affordable housing crisis, high cost of living, homelessness, gentrification, traffic congestion, and property crime (Brownstein 2017; Balk 2018a). The region's environment, a key feature, is experiencing increased environmental degradation (Wellman and others 2014), reflected in land use changes, biota reductions, and increases in chemical contamination (Puget Sound Partnership 2019). Just as degradation is human-driven, its impacts are experienced by humans. Increased strain and decline of the ecosystem means increased threats to human health and well-being (Wellman and others 2014; Biedenweg 2016). For example, increased toxic contaminants increases fish consumption risks, particularly among indigenous communities (Nicole 2013). Thus, while a 2017 poll among Puget Sound residents found that locals consider the region to be moving in a positive direction, residents noted that growth is a challenge and not entirely positive (Atlantic Media/Allstate 2017).

Puget Sound's growth may have social and environmental costs. These pressures have not gone unnoticed, as demonstrated by a protests, ballot measures, government actions, and civic initiatives seeking to address growth-related problems, including affordable housing and environmental degradation (Johnson 2016; Semuels 2018; Ollikainen 2019). The costs have also led to scapegoating, stereotyping, and finger-pointing toward the region's newcomers, particularly those employed by the region's technology industry and/or residents from California (McGreal 2018; Martichoux 2018). This nativist blame extends beyond Puget Sound to the larger Pacific Northwest (Wang 2015; May 2017) and is not new to the region (Johnson 1993; Smith 1989). This regionalism or nativism questions newcomers' roots, connections, and fealty to the region and its problems.

But, are the region's newcomers wholly detached, blasé, and uninterested in the region's place-based concerns? Does the region's growing nonnative-born population care about the region's natural environment or engage in proenvironmental behaviors? In this paper, we address these questions, among others, by examining the relationships among place attachment, environmental stewardship, and residency. Through the analysis of data derived from the region's first human well-being survey (Fleming and Biedenweg 2019), we highlight the status of place attachment and its relationship to residency length and engagement in stewardship activities with the intention of informing how newcomers are framed and included in regional ecosystem restoration communications, planning, and policy. We achieve this in the following sections by outlining our approach to place attachment, place attachment's role in environmental fields, Puget Sound's ever-changing context, our applied methods and analysis, and regionally relevant results and conclusions.

Place Attachment and Ecosystem Recovery

place attachment and sense of place

Place attachment has a plurality of definitions, emphases, theories, and applications (Altman and Low 1992; Manzo and Devine-Wright 2014; Williams 2014;

Smith 2018). This plurality stems from place attachment's interdisciplinarity, multidimensionality, and connections to sense of place (Altman and Low 1992; Smith 2018). Place attachment is considered a bond or connection between people and place, notably, but not exclusively, the natural environment (Manzo and Devine-Wright 2014; Smith 2018). Place attachment as a people-place bond strongly exemplifies and can be equated with the notion of "topophilia," or the affective ties people have with the physical environment, popularized by humanistic geographer Yi-Fu Tuan (1979). A strong or positive place attachment creates a sense of rootedness, belonging, or an intense feeling of being at home (Hay 1998; Haartsen and Stockdale 2018). It reflects and includes cognitive (e.g., beliefs, memory, or knowledge), emotional (e.g., feelings), and practice (e.g., actions) components (Anton and Lawrence 2014, 2016; Smith 2018) of the human experience.

We approach place attachment as dimension of sense of place (Figure 1). While place attachment emphasizes people-place bonds, sense of place refers to people's attachments and meanings associated with place, both individual or shared (Tuan 1979; Altman and Low 1992; Smith 2018). As Shmuel Shamai notes, "attachment to place is one of the components of sense of place," and any "specifications of concepts subsumed under sense of place include place attachment" (2018, 1350). Other dimensions include: place meaning, referring to the place-based descriptions that define a place and the imagery a place evokes (Williams 2014); place dependence, defined as the instrumental, functional, or beneficial people-place relationship that permits the achievement of needs and goals (Masterson and others 2017); and place identity, illustrated through place-based identification in relation to the physical environment (natural or built), landscapes, nation-states, cities, or even place names (Helleland 2012; Anton and Lawrence 2016; Trimbach 2016, 2019). As illustrated by Figure 1, each dimension

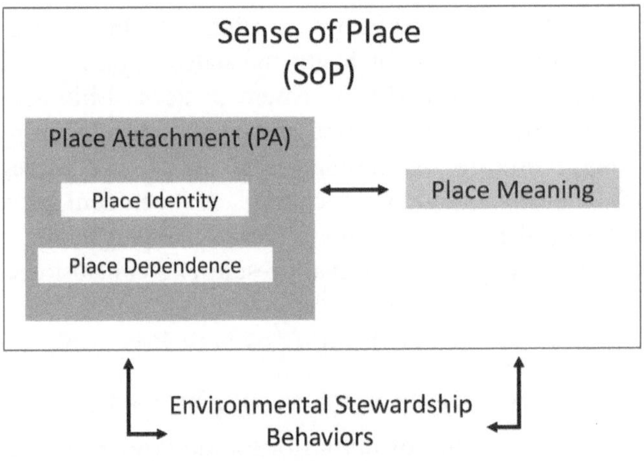

FIG. 1—Conceptual Framework (modified from Masterson and others 2017, 50).

interacts and/or informs one another (Masterson and others 2017). For example, place meaning informs place attachment, which itself includes place identity and place dependence, and informs place-based behaviors.

PLACE ATTACHMENT AND STEWARDSHIP

Sense of place and place attachment have expanded far beyond human geography (Smith 2018), most notably within environmental fields (Jorgensen and Stedman 2001; Manzo and Devine-Wright 2014; Masterson and others 2017; Kibler and others 2018; Ryfield and others 2019). This proliferation illustrates the "spatial turn" within the social sciences (Warf 2017), but also the wide application of place-based constructs and recognition of place within environmental management, planning, and policy, including ecosystem recovery (Biedenweg 2016; Masterson and others 2017; Kibler and others 2018; Ryfield and others 2019). This integration is also part of an ongoing shift towards social science integration in environmental and natural resource management (Robinson and others 2012; Biedenweg 2016). Sense of place and place attachment have emerged as exemplars of how social science constructs can support environmental-management efforts (Masterson and others 2017). This integration is partly founded on sense of place and place attachment being understood as informing and predicting proenvironmental behaviors (Jorgensen and Stedman 2001; Stedman 2002; Anton and Lawrence 2014; Junot and others 2018; Olstad 2018). This project is anchored in this understanding that place attachment contributes to place-based behaviors, like proenvironmental stewardship. This link posits that place attachment can impact responses to place change, policies, planning, or communities (Jorgensen and Stedman 2001; Anton and Lawrence 2014, 2016; Enqvist 2017; Masterson and others 2017; Junot and others 2018). For example, numerous scholars have shown that proenvironmental stewardship behaviors are linked, if not predicted, by a positive sense of place and/or place attachment (Jorgensen and Stedman 2001; Stedman 2002; Anton and Lawrence 2014, 2016; Junot and others 2018).

While this linkage is pronounced if not often assumed, this linkage has been challenged (Uzzell and others 2002; Williams 2014; Enqvist 2017). Due to a plurality of conceptualizations, operationalizations, and trajectories of sense of place and place attachment, there is conflicting evidence and understanding related to their applications and findings, including their impacts on behavior (Williams 2014). Additionally, some studies have shown that place attachment can lead to less proenvironmental behaviors (Uzzell and others 2002) as place attachment may not be the sole pathway to stewardship, but rather place meanings and people-place relations more broadly (Enqvist 2017). Regardless of such inconsistencies and critiques, both constructs offer practitioners the ability to assess, understand, and integrate place-based behaviors, values, feelings, identities, meanings, connections, and perceptions into environmental work. Our study contributes to this work by assessing whether or not place attachment influences stewardship in the Puget Sound region.

PLACE ATTACHMENT AND RESIDENCY

Like stewardship, place attachment is also understood and often assumed to be linked to, if not predicted by residency, specifically residency length (Anton and Lawrence 2014, 2016). This understanding equates residency length with a strong or positive place attachment. Interestingly, research has found that mobile, nonpermanent, or new residents have multiple, simultaneous attachments, identities, meanings, and dependencies to places (Ehrkamp and Leitner 2006; Gustafson 2014; Trimbach 2016; Morse and Mudgett 2018). Additionally, more recent research has shown that newly settled residents and even nonpermanent residents can develop a sense of place and place attachment to a place (Kelly and Hosking 2008; Watt 2009; Haartsen and Stockdale 2018; Morse and Mudgett 2018; Di Masso and others 2019).

For example, Kelly and Hosking (2008) found that nonpermanent residents can form a high level of place attachment. They noted that nonpermanent residents had been ignored by local decision makers based on the assumption that nonpermanent residents were not fully part of or connected to the place-based community; however, the authors argue that these attachments reflect positive people-place relationships that could be used to instill positive behaviors. Haartsen and Stockdale (2018) note that many newcomers are spatially and socially selective, allowing them to develop place attachment and sense of "s/elective belonging" to their new homes, while simultaneously allowing them the freedom to move (Haartsen and Stockdale 2018).

While newcomers and long-term residents may share a place attachment with a place, this is not to say that they are identical or founded on the same place meanings. For example, within the context of middle-class newcomers relocating to a mixed-income area (includes low-income residents), Watt (2009) found that newcomers were attached to the area; however, that attachment reflected a selective belonging in which they bonded with specific attributes or places within that area. Thus, a newcomer with a strong or similar place attachment to a long-term resident may imbue very different meanings to a place compared to a long-term resident with rootedness or rooted sense of place to the same area (Hay 1998). This understanding runs counter to the often-held interpretation that newcomers' sense of place is weaker, divergent from, or even threatening to places or shared sense of place among established residents (Creighton, Blatner, and Carroll 2008; Anton and Lawrence 2014). Overall, such findings support Di Masso and others' (2019) recent call to reimagine the relationship between mobility and place attachment "by recognizing that individuals and communities may have different types, valences and intensities of place attachments depending on varying and overlapping modes of interrelation between mobility and immobility" (131). Our study builds upon this shift to reimagine people-place relationships by examining the role residency.

While concurrent streams of research are being conducted on residency, place attachment, and stewardship, we are unaware of empirical studies that have explored these factors in unison. This study examines whether place attachment is a predictor of proenvironmental stewardship behaviors, and whether or not length of residency matters to place attachment and stewardship among Puget Sound residents.

Puget Sound Recovery Context

PUGET SOUND

Puget Sound is the largest estuary by volume in the United States (Figure 2) and a constituent part of the transboundary inland Salish Sea, that also includes the Strait of Juan de Fuca and Strait of Georgia. This fjordal estuarine system extends from Olympia in the south to Admiralty Inlet in the north at the Canadian border. Carved by glaciers approximately 10,000 years ago, Puget Sound encompasses 2,500 miles of shoreline and 2,800 square miles of inland marine waters (Puget Sound Partnership 2019), distinguished as four large marine basins, including Hood Canal, Main Basin, South Basin, and Whidbey

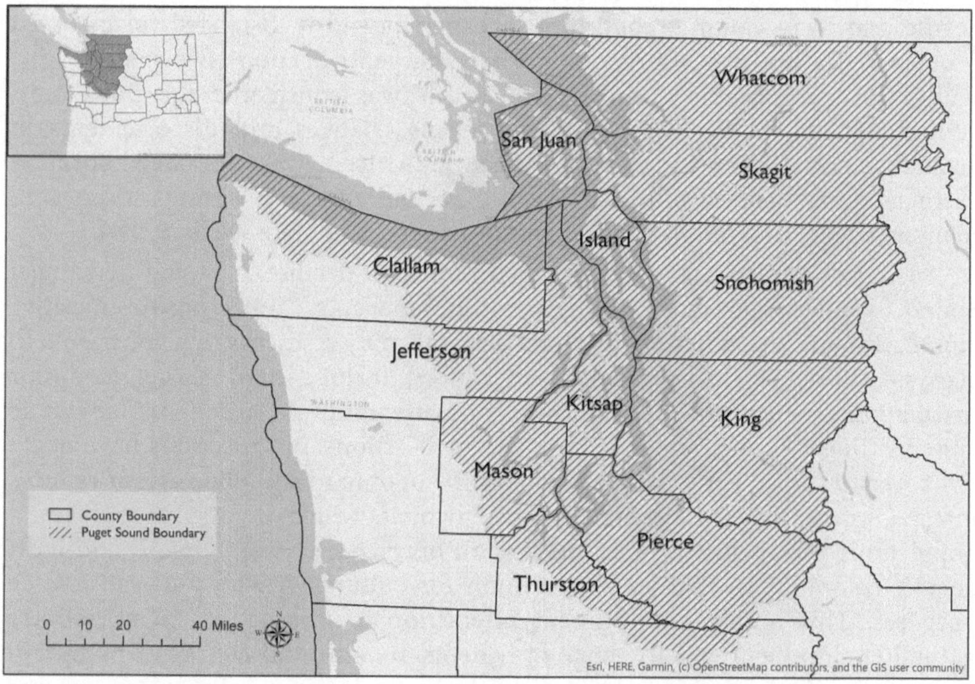

FIG. 2—Map of Puget Sound, Washington.

Basin (Puget Sound Institute 2015). An estimated 2,800 streams, ranging from small creeks to large rivers, flow into all four basins (Puget Sound Institute 2015).

The Puget Sound region is a dynamic place and includes rich indigenous cultures, diverse habitats, booming technology industries, charismatic orca species, and awe-inspiring landscapes. Ecologically, Puget Sound includes freshwater, terrestrial, nearshore, and estuarine habitats (Puget Sound Institute 2015). These habitats are integral to supporting a range of species, including salmon, kelp, oysters, eel grass, crab, geoduck, and herring that contribute to a wider marine food web that includes orca, seals, and humans (Wellman and others 2014). Puget Sound is considered a biodiversity hotspot as the region is home to 7,000 species, both terrestrial and aquatic (Wellman and others 2014).

Such biodiversity and ecological richness benefit the diverse communities that reside in region's twelve counties. Puget Sound's counties are home to 68 percent of Washington State's population, or a total of 4,779,172 out of 7,061,530, including over 20 federally recognized and unrecognized tribal communities (Puget Sound Institute 2015). Partly the result of economic growth, King, Pierce, Snohomish, and Kitsap counties gained over 500,000 new residents since 2010 and Seattle alone gained 16,900 new residents between 2018-2019 (Atlantic Media/Allstate 2017; Brownstein 2017; Puget Sound Regional Council 2019). As of 2018, less than half of Washington's residents (46 percent) were born in the state, with many (approximately 600,000) relocating from California (Savransky 2019).

Such growth is not without social consequences. Puget Sound, particularly Seattle and King County, continues to experience growth-related, place-based problems, like housing affordability crises and gentrification (Brownstein 2017; Balk 2018a). For example, as of 2019, Seattle was considered the third most gentrifying city in the United States, causing changes and displacements in Seattle's Central District, the city's historic African-American neighborhood (Scruggs 2018; Balk 2019). Such issues have led to some outmigration, with many residents moving to nearby counties and out of state (Balk 2018b).

Such growth also contributes to environmental decline. It is said that Puget Sound is suffering from a "death by a thousand cuts," reflecting the negative cumulative impacts of actions and inactions taken by residents that contribute to ecosystem decline (Ryan 2019). Such actions include hard shoreline armor installation whereas inactions include not attending to failing on-site septic systems (Puget Sound Partnership 2019). Such actions and inactions have negative reverberations. For example, the installation of hard shoreline armor reduces integral forage fish and juvenile salmon habitat, which in turn reduces their populations with negative repercussions on orca, recreational anglers, and indigenous communities that rely on salmon for cultural, health, and subsistence purposes. This human-environment interaction is striking, as Puget Sound's natural environment and its range of benefits are also pull factors of migration to the region (Atlantic Media/Allstate 2017; Brownstein 2017). In order to ensure

Puget Sound health and recovery, monitoring and planning actions are ongoing through the Puget Sound Partnership and wider ecosystem recovery community.

Puget Sound partnership

After efforts to address Puget Sound's decline, former Washington Governor Christine Gregoire, through the legislature, created the Puget Sound Partnership (Partnership) in 2007. The Partnership's key task is to lead and coordinate ecosystem restoration and protection efforts, working with a range partners including tribal, state, federal, and local governments. Although a nonregulatory agency, the Partnership collaborates with a range of partners to monitor and manage recovery. The Partnership is governed by an environmental governance regime known as the Management Conference, which includes the aforementioned partners, and overseeing bodies that include the Leadership Council, Ecosystem Coordination Board, and Science Panel, all of whom embody different roles and responsibilities. For example, the Science Panel is comprised of regional scientists and ensures the Partnership develops a science-based recovery plan for the region (Biedenweg and others 2017).

Vital Signs (indicators) are a key aspect of the Partnership's efforts. These indicators measure both biophysical attributes of ecosystem health and environmentally specific social indicators of human well-being (Biedenweg 2016; Biedenweg and others 2017). Biophysical indicators include abundance of chinook salmon, Pacific herring, and orca whales, for example, whereas human well-being indicators include average general public rating of their sense of place, and frequency with which the general public engages in environmental stewardship activities (Puget Sound Partnership 2019). This work contributes to regional research focused on sense of place in the region (Poe, Donatuto, and Satterfield Poe; Trimbach 2019), including paralleled indicator work that has highlighted the distinct sense of place among the region's indigenous population (Donatuto and others 2016). Unlike the biophysical indicators that are monitored and updated by external partners, many of the human well-being indicators are monitored by a regional general public human well-being survey conducted every two years to track change or recovery goal progress over time.

Methods and Analysis

Data

This study uses data from over 2,000 respondents to the 2018 Puget Sound Partnership human well-being survey to examine the connections among place attachment, stewardship behaviors, and length of residency within the Puget Sound's context of rapid demo-economic growth and concurrent environmental decline. The dataset also includes a variety of demographic factors in addition to these three variables.

Data were collected using a survey mailed to a random sample of 9,000 addresses, clustered for equal distribution across the twelve counties. The researchers opted for a clustered sample to obtain better geographic representation from urban and rural residents, rather than using a simple random sample where responses may have been biased toward urban populations. Single delivery point receptacles that service multiple residences, seasonal residences (e.g. summer homes), and vacant addresses were excluded from the initial sample. Of the original 9,000 addresses, surveys were only sent to 8,938 because 62 of the addresses were determined to be outside Washington.

The survey instrument was distributed using a modified Tailored Design Method (Dillman 2014) common in high "N" sociological and social-psychological studies. An initial notification postcard was sent to all addresses two weeks prior to the survey. Any postcards that came back as undeliverable were removed from subsequent mailings. A packet was then mailed to remaining addresses containing the paper survey, a letter of introduction, a prepaid return envelope, and a $2.00 bill incentive. The introduction letter included an URL allowing participants to take the survey online via Qualtrics. One week after survey packets were mailed, a reminder postcard was sent, which also included the URL. The online version of the survey attempted to mirror the paper instrument. An additional prompt in the sense of place section (I feel connected to the Puget Sound) was accidentally and only included in the online version. Results from this prompt were excluded. Paper (n=2,076) and online (n=246) survey data were entered into one database for a total sample size of 2,322. From the initial postcard mailing of 9,000, there were 677 undeliverable addresses and 62 addresses outside of Washington, resulting in an overall response rate of 28 percent.

VARIABLES

This study only uses the variables associated with place attachment (including sub-components of dependence and identity), stewardship, and length of residency from the survey. The survey questions were determined through a collaborative process involving Puget Sound communities and social scientists to identify the best available metrics from social scientific literature that simultaneously would be understood by the diverse population (Biedenweg and others 2017). Three questions were asked about frequency of individual participation in stewardship activities (see Table 1), built on the social psychology of engaging in long-term behaviors that has found that individuals do so if they are motivated by something they care about, either through moral obligation, personal needs, or perceived benefits (Measham and Barnett 2008). We did not provide examples of stewardship behavior, leaving this to the respondents' interpretation and focusing instead on the motivational factors. Response options were on a five-point frequency scale: (1) never, (2) rarely (1-4 times per year), (3) occasionally (once per month), (4) regularly (once per week), (5) frequently (almost

TABLE 1—MEAN AND STANDARD DEVIATION OF STEWARDSHIP BEHAVIOR AND PLACE ATTACHMENT METRICS IN THE PUGET SOUND

VARIABLES	MEAN(M)	STD. DEV. (SD)
Stewardship Behavior[1]		
Behaviors that you believe effectively benefitted the environment	3.50	1.30
Environmental behaviors that were personally meaningful to you	3.50	1.30
Environmental behaviors that you believe are needed by the community	3.30	1.30
Place Attachment[2]		
I am very attached to the natural environment in the Puget Sound region (PA 1)	6.10	1.10
I am proud to live in the Puget Sound region (PA 2)	6.30	1.10
Living in the Puget Sound region says a lot about who I am (PA 3)	5.50	1.40
Being able to engage in outdoor activities or cultural practices is important to my connection to the Puget Sound (PA 4)	5.90	1.30
I am mostly attached to parts of Puget Sound that are nearest to me (PA 5)	5.50	1.40
I could be satisfied living in other places outside the Puget Sound[3] (PA 6)	3.70	1.70

[1]Variables measured on a 5-point scale of level of engagement of 1 "Never", 2 "Rarely (1-4 time a year)", 3 "Occasionally (Once a month)", 4 "Regularly (Once a week)", to 5 "Frequently (Almost every day)"

[2]Variables measured on a 7-point scale from 1 "strongly disagree" to 7 "strongly agree"

every day), and included an option for "don't know." Answers of "don't know" were treated as missing.

Six questions were derived from social psychological literature (Jorgensen and Stedman 2001; Stedman 2002) and the collaborative process to measure place attachment (note: "PA" represents place attachment measures) (see Table 1). PA 1, PA 2, PA 5, and PA 6 are direct measures of place attachment. Question PA 3 measures place identity and question PA 4 measures place dependence. Respondents were asked the extent to which they agreed with PA statements on a bipolar scale from strongly disagree (1) to strongly agree (7) and included an option for "don't know." "Don't know" answers were treated as missing.

While few studies have found demographic predictors of social-psychological variables such as place attachment and environmental stewardship, we nevertheless wanted to control for demographic explanations. Demographic variables included age, sex, income, years of education, years lived in the Puget Sound, area, and race, and are used as controls. Age and income brackets matched census brackets. Sex and race were coded as dummy variables with white (0) and nonwhite (1), and male (0) and female (1) respectively. The survey included an "other" category for gender, but only three respondents chose this answer, providing an insufficient sample to separately analyze this group. Years of education ranged from 1 to 24+. Years lived in the Puget Sound was an open-ended question. Out of the total 2,322, respondents, 11.6 percent (269 persons) resided in the region from 0-5 years and 5.4 percent resided in the region from

6-10 years (125 persons), illustrating those who could be classified as "newcomers." Respondents could also choose rural, suburban, or urban for area.

DATA ANALYSIS

Excluding demographic variables, approximately 33 percent of surveys had at least one missing answer (including demographic variables, around 50 percent of individuals had missing data, mostly due to the income question). Missing response rates for place attachment, stewardship, and years of residence questions were between 1 percent and 11 percent. To address nonresponse, we preformed multiple imputation using the *Hmisc* package in R (Harrell and others 2019, function *aregImpute*). This method uses bootstrapping along with additive regression and predictive mean matching. Ten imputations were performed, predicted values were averaged across the ten imputations, and average values were substituted for missing values in the original data. Data were weighted for age to match age distributions from the 2010 census (the most recent census).

Cronbach alpha reliability coefficients were calculated for respondents' place attachment and stewardship behavior to test internal reliability of scales. This reliability analysis shows whether individual responses to the same topic are being answered with enough similarity that answers to all questions can be averaged to form an index. Place attachment, stewardship, and residency were mapped to display spatial patterns in the data using mean values within one-km grid cells. We did not use counties as our administrative units as they are quite large and do not show the diversity of responses.

Frequency statistics were also calculated and correlations between independent variables were tested. We used the *Survey* package in R (Lumley 2020, function *svyglm*) to perform ordinary least squares multiple regression analyses to determine the association of place attachment, years of residence, and demographics on stewardship behavior.

RESULTS

On the three questions pertaining to their stewardship behavior, individuals indicated they participated in stewardship activities between occasionally (once per month) and regularly (once per week). Average answers were very similar for the three questions; behaviors that individuals believed effectively benefitted the environment (M=3.50 on a scale of 1-5) were performed as frequently as environmental behaviors that were personally meaningful to individuals (M=3.50). Individuals indicated they performed environmental behaviors that they believed were needed by the community just slightly less frequently (M=3.30) (Table 1).

Answers to the place attachment questions varied more in comparison, with most individuals indicating a strong place attachment for three questions in particular. On average individuals agreed they were very attached to the region's natural environment (*M*=6.10 on a scale of 1-7), they were proud to live in the Puget Sound (*M*=6.30), felt responsible for the environment (*M*=6.10), and being able to engage in outdoor activities or cultural practices was important to their place-based connection (*M*=5.90) (Table 1). The other three questions differed, as individuals were between slight agreement and agreement about living in the Puget Sound region saying a lot about who they are (*M*=5.50) and being mostly attached to parts of Puget Sound that are nearest to them (*M*=5.50). Individuals were between neutral and slight disagreement that they could be satisfied living outside the Puget Sound (*M*=3.60, Table 1).

Cronbach alpha reliability coefficient was used to test internal consistency of stewardship behavior and place attachment indices. For example, if a segment of a population who regularly engage stewardship behavior due to personal meaning also say they occasionally perform stewardship behaviors due to environmental effectiveness, we can say there is consistency in how similar types of people engage the "stewardship" index, justifying the formation of an index rather than three individual stewardship variables. An alpha correlation coefficient above .65 is generally considered acceptable to justify creation of an index (Vaske 2008). Reliability coefficients showed high internal consistency for both variables: stewardship behaviors had a coefficient of .92 and place attachment questions showed lower reliability with an alpha of .73. Removing the items "I am mostly attached to parts of Puget Sound that are nearest to me" and "I could be satisfied living in other places outside the Puget Sound" from the place attachment index increased the reliability coefficient to .75. We removed those from the index and tested those questions separately for regression analyses.

Correlations between independent variables (all except stewardship) were tested, and all correlations were below .4, which is considered small (Vaske 2008). A variance inflation factor was also calculated between all independent variables, which showed no concerning multicollinearity between variables with factors ranging from 1.02-1.86. Partial correlation of place attachment and stewardship given years was .12 ($p<.001$). These figures all indicate acceptable levels to run regression analyses. To visually compare the similarities and differences between our main variables of interest, we mapped the mean values of participant responses from place attachment and stewardship behavior across the Puget Sound using one-km grid cells (Figure 3).

To compare the three different regression models with different combinations of place attachment questions, we used Akaike information criterion (AIC), a measure of the quality of statistical models where lower scores are better models (Akaike 1998). Each model contained years of residence, demographics (sex, area, education, income, race, and age), and place attachment as independent variables predicting the stewardship behavior index. The first model (1) included the index of PA variables, excluding PA 5 and PA 6, as they lowered the alpha of items. The second model (2) included the same index, as well as PA 5

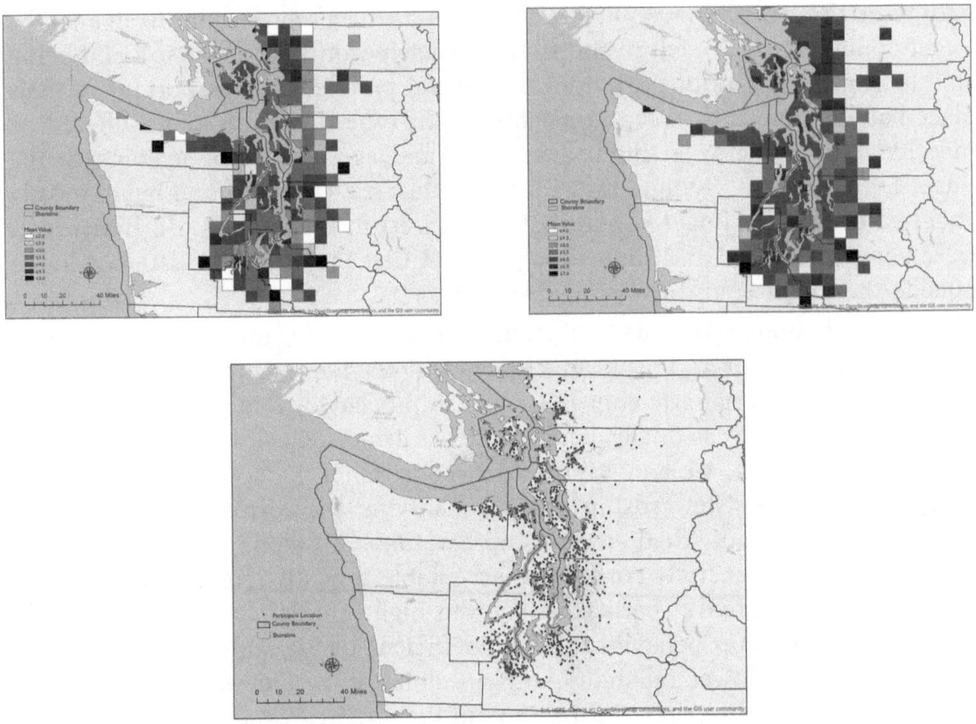

FIG. 3—Maps displaying mean values of variables using one-kilometer grid cells, and locations of participants. Map A shows mean values of the stewardship index variable. Map B shows mean values of the Place Attachment index variable. Map C shows individual point locations of participants.

TABLE 2—AIC COMPARISON FOR THREE DIFFERENT MODELS WHERE STEWARDSHIP BEHAVIOR IS THE DEPENDENT VARIABLE

MODEL	K	AIC
1	10	7109.08
2	12	7089.58
3	15	7001.95

and PA 6. The third model (3) used all place attachment questions separately rather than an index. The third model had the lowest AIC score and so was the main model selected (Table 2).

In the third model, place attachment, years of residency, and demographics explained 16 percent of the variation in responses to stewardship behaviors in the Puget Sound. Being a woman (b=-.28), and place attachment (PA 1) (b=.33), had the largest association with stewardship behavior. Being satisfied with living somewhere else (PA 7) (b=.08), and education (b=.04) also exhibited a

TABLE 3—REGRESSION ANALYSIS FOR PREDICTING STEWARDSHIP BEHAVIORS IN THE PUGET SOUND

	DEPENDENT VARIABLE: SOUND STEWARDSHIP INDEX [1]			
	B	*SEB*	*T*-VALUE	*P*-VALUE
PA 1[2]	0.33	0.04	7.38	<.001
PA 2[2]	0.00	0.05	0.01	.99
PA 3[2]	0.01	0.03	0.36	.72
PA 4[2]	0.03	0.04	0.85	.40
PA 5[2]	-0.03	0.03	-1.21	.22
PA 6[2]	0.08	0.02	3.71	<.001
Years Lived in the Puget Sound[3]	0.00	0.00	-0.04	.97
Sex[4]	-0.28	0.08	-3.77	<.001
Area[5]	0.03	0.04	0.59	.55
Education[6]	0.04	0.01	3.61	<.001
Income[7]	0.00	0.02	0.06	.96
Race[8]	0.03	0.09	0.37	.71
Age[9]	0.02	0.03	0.91	.36

[1]R^2 = 0.163; adjusted R^2 = 0.158
[2]Item coded from 1 "strongly disagree" to 7 "strongly agree"
[3]Item coded continuously
[4]Dummy variable where women are "0" and men are "1"
[5]Factor where urban is "1," suburban is "2," and rural is "3"
[6]Item coded continuously from 1 year of education to 24+ years
[7]Item coded from 1 "Less than $10,000" to 8 "Greater than $200,000"
[8]Dummy variable where white are "0" and nonwhite are "1"
[9]Item coded from 1 "18-24" to 7 "Greater than 75"

significant association with stewardship behaviors at the $p<.01$ level, but with smaller associations. All other variables, including length of residence, were not significantly associated with stewardship behaviors in the model (Table 3).

CONCLUSION AND DISCUSSION

The results of this study illustrate that length of residency does not meaningfully correlate to place attachment or stewardship behaviors. In fact, place attachment to the region's natural environment was high among all residents in Puget Sound. This strong place attachment was particularly pertinent for targeted place attachment statements (see Table 1), including PA 1 ("I am very attached to the natural environment in the Puget Sound region") and PA 6 ("I could be satisfied living other places outside the Puget Sound"). The results are also supported by the regression analysis (see Table 3), that conveys the length of residence again were not linked or predictive of stewardship behaviors. This strong place attachment reveals that Puget Sound's newcomers are not detached, blasé, or uninterested in local place-based concerns. On the contrary, the region's newcomers, like long-term residents, hold a positive sense of place and strong place attachment that likely facilitates or informs their shared place experiences, emotions, embodiments, and engagements. Thus, residents new and old

that they engage in stewardship, their actual engagement may widely vary, from minor-informal to major-formal stewardship actions. Additionally, while newcomers to the region may share a positive sense of place and strong place attachment to Puget Sound's natural environment with long-term residents, we cannot describe qualitative distinctions or variations due to the study's reliance on quantitative measures. As Gustafson notes, a reliance on quantitative empirical approaches limits nuance or conceptual complexity and "neglects important qualitative differences" (2014, 38). For example, while the survey results reflect residents' high place attachment to Puget Sound, our results lack any descriptive details about the shared or differing place meanings that inform that high place attachment among new and long-term residents. The benefit of this quantitative approach, however, was the ability to measure our three variables of interest across a large representative sample of our population in order to run statistical tests of significance. More qualitative studies, while better at solicitating full understanding of place meanings, are less effective at broadly characterizing perceived sense of place as determined by place attachment and place identity across a spectra of demographic variation.

In conclusion, Puget Sound belongs to all of its residents and the region's problems are shared by and are the responsibility of all residents. While specific communities may inequitably suffer consequences of specific aspects of degradation, particularly indigenous peoples with a distinct historical heritage and subsistence users of the region's natural resources, the Puget Sound's declining ecosystem health impacts all residents. Although this study showed that residency length did not matter to residents' place attachment, and stewardship behaviors, it does not negate other place-based issues that are associated with growing populations, including gentrification, traffic congestion, housing affordability, and increased impervious surfaces, among others. These changes are fundamentally shared and not inherently or solely the result of conflated individual newcomer mindsets or actions. Such changes require recognizing the antinewcomer or sedentary biases (Di Masso and others 2019) embedded within media, community, or political narratives of the region's newest neighbors. This recognition can help deconflate and acknowledge the potential role this large community can play in proenvironmental stewardship. Such changes also necessitate planning and policy shifts, such as better directly engaging and integrating stakeholder communities, like new residents or technology company employees, into environmental decision making or action-oriented campaigns, like those facilitated by the Puget Sound Partnership. By better integrating and capitalizing on the strong place attachment and proenvironmental stewardship dispositions of all residents, the region's environmental leaders and decision makers can spark collective proenvironmental stewardship actions (Kibler and others 2018; Plummer and others 2020) that are more inclusive and effective at addressing the health and recovery of Puget Sound.

Funding

Survey implementation and data tabulation were supported by the Puget Sound Partnership through funds appropriated by the State of Washington State *under*

USEPA assistant agreement PC-01J32201. Salaries and benefits for personnel to implement the survey and run statistical analyses were funded by EPA STAR early career award #83694601; Environmental Protection Agency [EPA Star / #83694601,PC-01J32201].

ORCID

David J. Trimbach http://orcid.org/0000-0003-1402-4137

References

Akaike, H. 1998. Information Theory and an Extension of the Maximum Likelihood Principle. In *Selected Papers of Hirotugu Akaike*, edited by E. Parzen, K. Tanabe, and G. Kitagawa, 199–213. New York: Springer.

Alcock, I., M. P. White, S. Pahl, R. Duarte-Davidson, and L. E. Fleming. 2020. Associations Between Pro-Environmental Behaviour and Neighbourhood Nature, Nature Visit Frequency and Nature Appreciation: Evidence from a Nationally Representative Survey in England. *Environment International* 136: 105441. doi:10.1016/j.envint.2019.105441

Altman, I., and S. M. Low. 1992. Place Attachment: A Conceptual Inquiry. In *Place Attachment*, edited by I. Altman and S. M. Low, 1–12. New York: Plenum Press.

Anton, C. E., and C. Lawrence. 2014. Home Is Where the Heart Is: The Effect of Residence on Place Attachment and Community Participation. *Journal of Environmental Psychology* 40: 451–461.

_____, and _____. 2016. The Relationship between Place Attachment, Theory of Planned Behavior and Residents' Response to Place Change. *Journal of Environmental Psychology* 47: 145–154.

Atlantic Media/Allstate. 2017. The Allstate Renewal Project: Seattle Poll. [http://www.therenewalproject.com/wp-content/uploads/2017/11/Topline_Allstate_Renewal-Project-Seattle_09.29.17-1.pdf].

Balk, G. 2018a. Fed Up with Seattle? Data Show Where People Are Moving. *The Seattle Times*. 3 May.

_____ 2018b. 114,000 More People: Seattle's Now Decade's Fastest-growing Big City in All of U.S. *The Seattle Times*. 24 May.

_____ 2019. Seattle Is the Third Most Gentrifying U.S. City-but that Might Not Be as Bad as You Think, Study Finds. *The Seattle Times*. 24 July.

Beeco, J. A., and G. Brown. 2013. Integrating Space, Spatial Tools, and Spatial Analysis into the Human Dimensions of Parks and Outdoor Recreation. *Applied Geography* 38 (1): 76–85.

Berkson, B., and L. Dubois. 2019. Sound Behavior Index-2019 Survey Report. Bellingham: Western Washington University Center for Economic and Business Research.

Biedenweg, K. 2016. A Comparative Study of Human Well-Being Indicators across Three Puget Sound Regions. *Society & Natural Resources* 30 (3): 362–376.

_____., H. Harguth, and K. Stiles. 2017. The Science and Politics of Human Well-being: A Case Study in Cocreating Indicators for Puget Sound Restoration. *Ecology and Society* 22 (3): 11.

Brownstein, R. 2017. Can Seattle Handle Its Own Growth? *The Atlantic*. 16 November.

Creighton, J. H., K. A. Blatner, and M. Carroll. 2008. People, Place, and Politics: The Role of Place Attachment and Conflict in Forest Communities. *Western Journal of Applied Forestry* 23 (4): 232–235.

Di Masso, A., D. R. Williams, C. M. Raymond, M. Buchecker, B. Degenhardt, P. Devine-Wright, A. Hertzog, M. Lewicka, L. Manzo, A. Shahrad, R. Stedman, L. Verbrugge, and T. von Wirth. 2019. Between Fixities and Flows: Navigating Place Attachments in an Increasingly Mobile World. *Journal of Environmental Psychology* 61: 125–133.

Dillman, D. A., J. D. Smyth, and L. M. Christian. 2008. 3rd ed. *Internet, Mail, and Mixed-Mode Surveys: The Tailored Design Method*. Hoboken, N.J.: Wiley.

Donatuto, J., L. Cambell, and R. Gregory. 2016. Developing Responsive Indicators of Indigenous Community Health. *International Journal of Environmental Research and Health* 16 (9): 899.

Enqvist, J. 2017. Stewardship in an Urban World: Civic Engagement and Human-nature Relations in the Anthropocene. PhD diss., Stockholm University.

Fleming, W., and K. Biedenweg. 2019. Visualizing Human Well-being in the Puget Sound. Olympia, Wash.: Puget Sound Partnership

Florida, R. 2019. Why U.S. Tech Inventors Are So Highly Clustered. *City Lab.* [https://www.citylab.com/life/2019/10/technology-inventions-where-work-patents-agglomeration-data/599089/].

Gustafson, P. 2014. Place Attachment in the Age of Mobility. In *Place Attachment: Advances in Theory, Methods and Applications*, edited by L. C. Manzo and P. Devine-Wright, 37–48. London: Routledge.

Haarsten, T., and A. Stockdale. 2018. S/elective Belonging: How Rural Newcomer Families with Children Become Stayers. *Population, Space and Place* 24: e2137.

Hay, R. 1998. A Rooted Sense of Place in Cross-Cultural Perspective. *The Canadian Geographer* 42 (3): 245–266.

Helleland, B. 2012. Place Names and Identities. *Oslo Studies in Language* 4 (2): 95–116.

Holder, S. 2019. Amazon's Slow Retreat from Seattle. *City Lab.* [https://www.citylab.com/equity/2019/04/amazon-seattle-headquarters-tech-jobs-bellevue-crystal-city/586549/].

Johnson, M. S. 1993. *The Second Gold Rush: Oakland and the East Bay in World War II.* Berkeley: University of California Press.

Johnson, P. 2016. Are Washingtonians About to Vote Down a Chance to Fight Climate Change? *The Atlantic.* 7 November.

Jorgensen, B., and R. Stedman. 2001. Sense of Place as an Attitude: Lakeshore Owners' Attitude toward Their Properties. *Journal of Environmental Psychology* 21: 233–248.

Junot, A., Y. Paquet, and F. Fenouillet. 2018. Place Attachment Influence on Human Well-being and General Pro-environmental Behaviors. *Journal of Theoretical Social Psychology* 2: 49–57.

Kelly, G. and K. Hosking 2008. Nonpermanent Residents, Place Attachment, and "Sea Change" Communities. *Environment and Behavior* 40 (4): 575–594.

Kibler, K. M., G. S. Cook, L. G. Chambers, M. Donnelly, T. L. Hawthorne, F. I. Rivera, and L. Walters. 2018. Integrating Sense of Place into Ecosystem Restoration: A Novel Approach to Achieve Synergistic Social-ecological Impact. *Ecology and Society* 23 (4): 25.

Leitner, H., and P. Ehrkamp. 2006. Transnationalism and Migrants' Imagining of Citizenship. *Environment and Planning A* 38: 1615–1632.

Lumley, T. (2020). *Survey: Analysis of Complex Survey Samples (4.0) [Computer software].* https://CRAN.R-project.org/package=survey

Maki, A., A. R. Carrico, K. T. Raimi, H. B. Truelove, B. Araujo, and K. L. Yeung. 2019. Meta-analysis of Pro-environmental Behavior Spillover. *Nature Sustainability* 2: 307–315.

Manzo, L. C., and P. Devine-Wright. 2014. Introduction. In *Place Attachment: Advances in Theory, Methods and Applications*, edited by L. C. Manzo and P. Devine-Wright, 1–7. London: Routledge.

Martichoux, A. 2018. 'They Do Not Like Californians': How the Pacific Northwest Is Treating Transplants. *SFGATE.* 1 August. [https://www.sfgate.com/expensive-san-francisco/article/move-bay-area-to-pacific-northwest-oregon-boise-13120449.php].

Masterson, V. A., R. C. Stedman, J. Engvist, M. Tengö, M. Giusti, D. Wahl, and U. Svedin. 2017. The Contribution of Sense of Place to Social-ecological Systems Research: A Review and Research Agenda. *Ecology & Society* 22 (1): 49.

May, P. 2017. 'Go back to California' Painted on Prius in Portland. *The Mercury News.* 3 July.

McGreal, C. 2018. Is Bezos Holding Seattle Hostage? The Cost of Being Amazon's Home. *The Guardian.* 4 July.

Measham, T. G., and G. B. Barnett. 2008. Environmental Volunteering: Motivations, Modes and Outcomes. *Australian Geographer* 39 (4): 537–552.

Morrill, R., and P. Sommers. 2005. Seattle as a Digital City: Unexpected or Inevitable? *Canadian Journal of Regional Science* 28 (2): 349–368.

Morse, C. E., and J. Mudgett. 2018. Happy to Be Home: Place-Based Attachments, Family Ties, and Mobility among Rural Stayers. *The Professional Geographer* 70 (2): 261–269.

Nicole, W. 2013. Meeting the Needs of the People: Fish Consumption Rates in the Pacific Northwest. *Environmental Health Perspectives* 121 (11–12): A334–A339.

Niemiec, R. M., R. Willer, N. M. Ardoin, and F. K. Brewer. 2019. Motivating Landowners to Recruit Neighbors for Private Land Conservation. *Conservation Biology* 33 (4): 930–941.

OFM [Office of Financial Management]. 2017. County and City Data. Olympia: State of Washington. [https://www.ofm.wa.gov/washington-data-research/county-and-city-data].

Ollikainen, R. 2019. Orca Task Force Hears about Whale Watching, Dam Breaching. *Peninsula Daily News.* 11 September.

Olstad, T. 2018. Visitor Perception, Place Attachment, and Wilderness Management in the Adirondack High Peaks. In *Explorations in Place Attachment*, edited by J. S. Smith, 133–148. London: Routledge.
Plummer, R., J. Baird, S. Farhad, and S. Witkowski. 2020. How Do Biosphere Stewards Actively Shape Trajectories of Social-ecological Change? *Journal of Environmental Management* 261: 110–139.
Poe, M., J., Donatuto, and T. Satterfield. 2016. "Sense of Place": Human Wellbeing Considerations for Ecological Restoration in Puget Sound. *Coastal Management* 44 (5): 409–426.
Puget Sound Institute. 2015. Puget Sound Factbook. Tacoma: University of Washington Tacoma.
Puget Sound Partnership. 2019. State of the Sound Report. Olympia: State of Washington.
Puget Sound Regional Council. 2019. Region Adding 188 People a Day. 28 June. [https://www.psrc.org/whats-happening/blog/region-adding-188-people-day].
Robinson, P., K. Genskow, B. Shaw, and R. Shepard. 2012. Barriers and Opportunities for Integrating Social Science into Natural Resource Management: Lessons from National Estuarine Research Reserves. *Environmental Management* 50: 998–1011.
Ryan, J. 2019. Homeowners Keep Building Walls around Puget Sound. Biologists Are Taking out More. *KUOW*. 25 January. [https://www.kuow.org/stories/homeowners-keep-building-walls-around-puget-sound-but-biologists-are-taking-out-more].
Ryfield, F., D. Cabana, J. Brannigan, and T. Crowe. 2019. Conceptualizing 'Sense of Place' in Cultural Ecosystem Services: A Framework for Interdisciplinary Research. *Ecosystem Services* 36: 100907.
Savranksy, B. 2019. Most People Who Live in Washington Weren't Born Here. They're from … *Komo News*. 7 November.
Scruggs, G. 2018. As Seattle Gentrifies, Blacks No Longer Feel Welcome. *Reuters*. 22 July.
Semuels, A. 2018. How Amazon Helped Kill a Seattle Tax on Business. *The Atlantic*. 13 June.
Shamai, S. 2018. Measuring Negative Sense of Place: Israeli Settlers' Forced Migration. *Geojournal* 83: 1349–1359.
Smith, J. S. 2018. Introduction: Putting PLACE Back in Place Attachment Research, In *Explorations in Place Attachment*, edited by J. S. Smith, 1–16. London: Routledge.
Smith, W. 1989. Booming Seattle Tells Hip Californians to Just Stay Away. *Chicago Tribune*. 19 September
Stedman, R. C. 2002. Toward a Social Psychology of Place: Predicting Behavior from Place-based Cognitions, Attitude, and Identity. *Environment and Behavior* 34 (5): 561–581.
Stedman, R. C., and M. L. Ingalls. 2014. Topophilia, Biophilia and Greening in the Red Zone. In *Greening in the Red Zone: Disaster, Resilience and Community Greening*, edited by K. G. Tidball and M. E. Krasny, 129–144. Dordrecht, Netherlands: Springer.
Trimbach, D. J. 2016. Understanding Narva & Identity: Local Reflections from Narva's Russian speakers. *Baltic Worlds* 1-2: 4–12.
_____ 2019. Salish Sea Survey. Eastsound, Washington: The SeaDoc Society, University of California-Davis.
Tuan, Y. F. 1979. Space and Place: Humanistic Perspective. In *Philosophy in Geography*, edited by S. Gale and G. Olsson, 387–427. Dordrecht, Netherlands: Springer.
Uzzell, D., E. Pol, and D. Badenas. 2002. Place Identification, Social Cohesion, and Environmental Sustainability. *Environment and Behavior* 34 (1): 26–53.
Vaske, J. J. 2008. *Survey Research and Analysis: Applications in Parks, Recreation and Human Dimensions*. State College, Pennsylvania: Venture Publishing.
Wang, A. X. 2015. Portlanders Are Using Stickers to Ke ep Californians Out. *The Atlantic*. 5 September.
Warf, B. 2017. Spatial Turn. In *The Wiley-Blackwell Encyclopedia of Social Theory*, edited by B. S. Turner, K-S Chang, C. F. Epstein, P. Kivisto, J. M. Ryan, and W. Outhwaite, 1–3. Chichester, U.K.: John Wiley.
Watt, P. 2009. Living in an Oasis: Middle-class Disaffiliation and Selective Belonging in an English Suburb. *Environment and Planning A* 41: 2874–2892.
Wellman, K. F., K. Biedenweg, and K. Wolf. 2014. Social Sciences in Puget Sound Recovery. *Coastal Management* 42 (4): 298–307.
Williams, D. R. 2014. Making Sense of 'Place': Reflections on Pluralism and Positionality in Place Research. *Landscape and Urban Planning* 131: 74–82.

"TIED TO THE LAND": PIPELINES, PLAINS AND PLACE ATTACHMENT

CHRISTINA E. DANDO

ABSTRACT. Since first proposed, the Keystone XL and Dakota Access pipelines have been extensively covered by the media, shaping readers' perceptions of the pipelines, as well as perceptions of the places and peoples impacted by them. Using critical discourse analysis, this paper examines the media coverage, their Plains descriptions, and expressions of place attachment. Through the media's use of "place talk," it presents a hybrid Plains: placeless, yet with a strongly place-attached population who are "tied to the land." As conflicts over environmental and energy projects become increasingly contentious, place and place attachment are important for understanding the conflicts and potentially mobilizing resistance. *Keywords: climate change, discourse analysis, Great Plains, pipelines, place attachment, water.*

In April 2017, Jim Carlson, a Nebraska farmer, was quoted in a *New York Times* article on the Keystone XL pipeline: "It's the people who are tied to the land ... who understand what's at risk" (Smith 2017b). This article and others like it on the Keystone XL, as well as the Dakota Access Pipeline, shaped readers' understandings of peoples, places, and events (Figure 1).

The Keystone XL (KXL) pipeline was proposed in 2008 by TransCanada (now TC Energy), a Calgary company, to run from Alberta's Athabasca tar sands to Texas refineries (Hodges and Stocking 2016, 225).[1] Supporters say it will safely transport needed oil and offer employment and tax benefits (Schulte 2011). KXL has been resisted generally over climate change concerns and in Nebraska over potential Ogallala Aquifer contamination and the use of eminent domain to forcibly cross private property (Pierre-Louis 2017). One of President Joseph Biden's first acts after his inauguration was to revoke its federal permit (Gillies 2021). TC Energy announced as this article was going to press that it was terminating the project (Davenport 2021).

The Dakota Access Pipeline (DAPL) was proposed in 2014, to run from North Dakota oil fields to Illinois refineries, ports, and facilities (Meyer 2017). Energy Transfer Partners L.P. (ETP), a Texas company, received construction approval from the states it would cross and from the Army Corps of Engineers to cross Lake Oahe (a Missouri River reservoir) (McMaken 2016). Supporters say it will provide a safe domestic energy source and offer employment and tax benefits (Miller 2017). DAPL has been opposed by Native Americans over the disturbance of sacred sites, the threat to the Standing Rock Indian Reservation water supply, and climate change, supported by a coalition of environmental

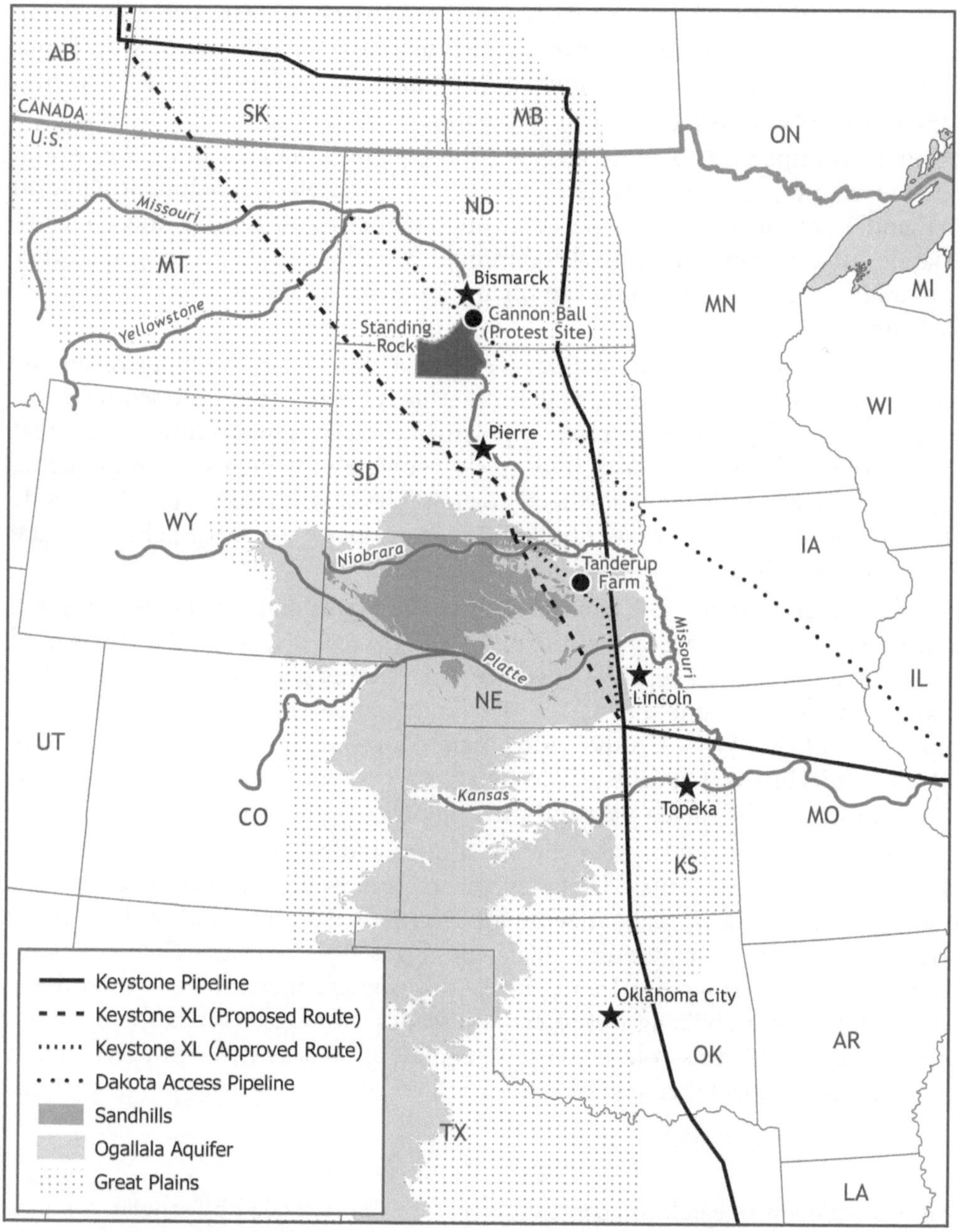

FIG. 1—Map of the Great Plains. Cartography by Paul Hunt.

justice and climate change groups (Sullivan 2016). At peak protests in late August 2016, an estimated 7000 protestors were at "water protector" camps (Gass 2016a).[2] Police and ETP-hired guards clashed violently with protestors in fall 2016 (Sullivan 2016). The pipeline was completed in April 2017, but legal challenges continue. In January 2021, a U.S. Court of Appeals upheld a ruling revoking the easement that allowed DAPL to cross beneath the Missouri, but permitting the oil flow to continue. With Biden's cancellation of KXL comes calls to also shut down DAPL (Forliti 2021).

Place is important to those resisting the pipelines, such as Jim Carlson and others trying to stop the pipelines. Sense of place and place attachment ("tied to the land") are but two sides of a coin, one attempting to capture those elusive qualities of a location that give it its unique character in human eyes, the other expressing the emotional bond humans have to locations (Smith 2017a, 2; Williams and Stewart 1998).

Place also matters to the media. News coverage of any event provides the who, what, where, when, why, and how, usually in the lede or opening paragraph. "Where" is a location, often supplemented with a description. The Great Plains are a challenging landscape for many people off the Plains to comprehend, often perceived as being "nowhere" or placeless, with little apparent value (Dando 2009). Did journalists provide a sense of place for the pipelines on the Plains? Or did they reinforce a sense of placelessness?

This paper considers the construction of the Plains and expressions of place attachment in media coverage of the pipelines. Critical discourse analysis is used to examine the media's pipeline coverage, the reasons given for resistance, their Plains descriptions, and the use of place attachment in their framing of the pipeline conflicts. What do the individuals quoted in the articles say they are attached to? How is it articulated? Why are they opposed to the pipelines? And how and why is the media utilizing these quotes in their articles?

This paper contributes to a better understanding of the media's role in shaping public perceptions of peoples, places, and events, as well as to a fuller understanding of how place attachment is articulated and employed by those "tied to the land." It may seem cliché to suggest "place matters," but it does. Conflicts over environmental and energy projects are increasingly contentious, with individuals mobilizing to protect landscapes they are attached to. We must understand the media's role in these struggles, as well as their employment of place and place attachment, as land use conflicts play out in these public forums.

Overview of Place Attachment

Place-attachment research in geography began in the 1970s, led by scholars such as Anne Buttimer (1980), Yi-Fu Tuan (1980), and Edward Relph (1976). Relph is often recognized for advancing our understanding of the emotions tied to place, from deep attachment (such as home) to placelessness (such as a strip mall) (Relph 1976, 6). A human bond with a location impacts thoughts, emotions, and connections of and to places, with place bonds serving important human purposes.

Leila Scannell and Robert Gifford propose a conceptual framework for considering place attachment, focusing on the attached actors, the psychological process of attachment, and the object of attachment (Scannell and Gifford 2010, 2). Scannell and Gifford echo journalisms' five w's and an h, emphasizing the importance of providing essential facts on place attachment. Scannell and Gifford's framework is important for considering place-attachment functions for humans: for survival and security, for goal support, continuity, and for identity and belongingness (2010, 5). Humans depend on their environments for survival; where humans find these necessities, they tend to cling, developing strong ties. Environments can be appreciated for providing space for valued activities, such as the space or resources needed to attain personal goals (for example, a business opportunity) or an environment for "self-reflection, problem-solving, and stress relief" (for example, a park) (Scannell and Gifford 2010, 6). Place attachment is often tied to identity, can be key to maintaining its continuity (reconnecting with their pasts or with cultural or religious identities), and can be shared for providing a sense of belonging to a larger community.

The media contributes to place attachment and to human understanding of places. In creating or reinforcing particular perspectives, the media can have positive as well as negative impacts on human place perception. Relph suggests the media contributes to placelessness by promoting uniform landscapes and cultures, providing "no roots, no sense of belonging to a place" (Relph 1976, 61, 90–92).

Since Relph's original work, research has addressed the intersection of place-attachment, media, and environmental issues (such as Shriver and Peaden 2009; Devine-Wright 2012) as well as specifically climate change (Devine-Wright 2013; Scannell and Gifford 2013; Upham et al. 2018). One significant research area is the threat to place and place attachment by forces such as economic development or climate change, with the greatest threat being place destruction (Adger et al. 2011, 3; Smith 2017a, 10). Douglas Porteous and Sandra Smith coined the term "domicide" to capture the intentional destruction of home "for political and/or corporate interests" (Nowicki 2014, 785; Porteous and Smith 2001; Porteous 1989). Mel Nowicki extends the concept, including " … the ways in which successful resistances [to domicide] are enabled from the ground up, both through highly visible resistances via public protest and petitioning and equally through banal performances of the everyday that aid and abet the damage domicide causes" (Nowicki 2014, 786).

A related research thread focuses on "disruptions" to place attachment. Patrick Devine-Wright considers the social and psychological aspects of new energy infrastructure siting, and how such projects often pit national or global benefits against local concerns. Wright proposes reconceiving NIMBY (not-in-my-backyard) responses "as place-protective actions … founded upon processes of place attachment and place identity" (Devine-Wright 2009, 432). Local inhabitants often perceive such projects as being " … 'imposed' upon places by companies or state organization without genuine public engagement" (Devine-Wright 2009, 434).

When a location is threatened, those identifying with the place may mobilize to save it (Cheng et al. 2013, 1170; Carrus et al. 2014, 160; see also Scannell and

Gifford 2013). The stronger the fight, the more protective they can be, and the more attached they can become (Tonge et al. 2015, 730). Place-protective actions can mobilize citizens removed from the immediate vicinity. Tyra Olstad considered how personal experiences may be used to rally the public (Olstad 2013, 105). Olstad explains that hearing of others' place attachments resonates with readers' own experiences and can deepen understanding and respect. "Imagined communities" of individuals linked by place attachment are formed and the media reinforces the linkages and associations (Devine-Wright 2013, 63; Tonge et al. 2015, 740).

These acts of place protection are important for the reciprocal relationship that exists between humans and the places they care about: "Maintaining the character of a place requires care from actors. Places, if they change their character, can then cease to sustain those who inhabit them, as the 'secure space' that makes embedded agency possible decays. With the loss of this secure space, the lived futures of its inhabitants may be undermined" (Groves 2015, 857). Moreover, as individual and community identity can be linked to place, threats to place are also threats to identity (Brown and Perkins 1992, 280).

Broadly, Relph's work is important in considering the Plains, a location where residents often express a deep sense of place attachment, yet outsiders frequently describe it as placeless. Native peoples have occupied the Plains for thousands of years, but little scholarly research has considered their sense of place or place attachment despite their clear embeddedness in specific landscapes (Estes 2019, 70). Early Europeans and Americans to the Plains projected onto its surface their expectations, be it a garden or a desert, passage or barrier (Allen 1985; Allen 1993). As the Great Plains were colonized, Europeans and Americans adjusted their gaze to the new land and came to associate it with home (Dando 2008). Over time and overcoming hardships, these settler-colonizers, now Plains residents, developed "an unyielding connection to place" (Smith and McAlister 2015, 183).

Methodology

The Lexis-Nexis/NexisUni database was utilized, as well as Internet searches, to locate pipeline articles set on the Plains, particularly newspaper and magazine articles published outside the Plains, and as straight or business news or feature articles (Table 1) (Budinksy and Bryant 2013, 211–212).[3] Such media sources are important for informing and shaping public opinion. Altogether, 171 articles were located, published between 2008 and 2018 (106 on the KXL and 65 on the DAP).[4]

Rudimentary content analysis was conducted on the articles, using a grounded theory approach (Reynolds 2018, 53–54).[5] Analysis considered the types of stories represented, what made the material "newsworthy," the arguments made, the language used, who was named and who was quoted, the conversations, and the

Table 1—Media sources used in this study

KEYSTONE XL	DAKOTA ACCESS
Bloomberg.com	Agweek
CBS News	The Atlantic
Christian Science Monitor	Christian Science Monitor
Ecologist	Esquire
E: the Environmental Magazine	Fortune
Esquire	The Hill
farmaid.org	Huffington Post
Forbes	The Humanist
Fortune	Los Angeles Times
Grist	Mother Jones
Huffington Post	Ms. Magazine
In These Times	The Nation
Inside Climate News	National Geographic
Los Angeles Times	National Public Radio
Mother Jones	Newsweek
The Nation	New York Times
National Geographic	pbs.org
National Review	Popular Science
The New Yorker	The Progressive
New York Times	Rolling Stone
Popular Science	Star-Tribune
PRI	St. Louis Post-Dispatch
The Progressive	Time
Reuters	Utne Reader
Rolling Stone	Wall Street Journal
Scientific American	Washington Post
Time	
USA Today	
US News & World Report	
Wall Street Journal	
Washington Post	

landscapes depicted and in what way, seeking to expose and explore the "persistent, politically relevant patterns" (Entman 2007, 164). Periodically, coding sheets were revised to reflect the topics surfacing, and the articles reanalyzed using the revised coding sheets. Separate coding sheets were created for the KXL and DAP to reflect the different conversations taking place (Table 2). From the coding process, codes were grouped under umbrella themes representing "the dominant discursive formations of the stories examined" (Reynolds 2018, 61). From this process, major themes employed by the media were identified (Table 3). Once major themes were identified, critical discourse analysis was utilized.

Table 2—Coding differences that emerged from the media analysis

KEYSTONE XL	COMMON CODES	DAKOTA ACCESS
High plains	Plains as vast	Camps
Heartland	Plains as flat	Tipis
Breadbasket	Prairie	Sacred space
Family farm	Grasslands	Pray*
Sandhills	Water	Burial/graves
Sandhills as delicate, fragile, vulnerable or sensitive	Irrigation	Spirit*
Ogallala Aquifer	Drinking water	Prophecy
Climate change/global warming	Water supply	Black Snake
Tar sands	Missouri River	Sovereignty*
Oil sands	Safety	Nation
Negatives of tar/oil sands	Jobs	Treaty*
Safety	Energy security	Treaty rights
Jobs	Threat of oil spills	Water protectors
Energy security	Account of actual oil spills	Movement
Landowner	Pipeline breaks	Solidarity
Private property	Farm* or Fields	Allies
"in the national interest"/"in the public interest"	Ranch* or grazing land*	Conflict
Reroute	Rural	Defense
	Reservation	Occupy*
	Cowboys	Flashpoint
	Indians	Front*
	Bison/buffalo	Resist*
	Frontier	Violent
	Citizen/citizenship/rights	Disorderly
	Easement	Trespass
	Eminent domain	Unlawful
	Environmentalists	Stand-off/Taking a stand/Last stand
	Environmental groups	
	Protest*	
	Demonstrate*	
	Oppose*	
	Activists	
	Block*	
	Arrest	
	Civil disobedience	
	Nonviolent	
	Peace*	
	Fight*	
	Battle*	
	War*/act of war	

Notes: An asterisk behind a word indicates coding included variations of the stem. For example: demonstrate* would include demonstrate, demonstration and demonstrator.

TABLE 3—MAJOR THEMES IN MEDIA NEWS COVERAGE, IN ORDER OF FREQUENCY. NUMBERS INDICATE APPROXIMATE NUMBERS OF FREQUENCY ACROSS THE SAMPLE

KEYSTONE XL		DAKOTA ACCESS	
Positive economic aspects (jobs/safety/energy security/less dependence)	191	Language of conflict or contestation (fight*/battle*/resist*/protest*/demonstrate*)	695
Ogallala Aquifer	188	Language of sacred/spiritual (sacred space/ceremonies/pray*/spirit*/prophecy*/Black Snake)	159
Oil spills/leaks/breaks (potential and actual)	185		
Tar sands	183	Language of rights & sovereignty (sovereignty*/rights*/nation)	74
		Water protector/s	73
Language of conflict or contestation (fight*/battle*/resist*/protest*/demonstrate*)	159	Treaties (treaty*)	68
		Water supply/drinking water	66
		Oil spills/leaks/breaks (potential and actual)	52
Sandhills as delicate (delicate/vulnerable/sensitive/fragile)	156	Positive economic aspects (jobs/safety/energy security/less dependence)	37
Oil sands	117		
Water supply/drinking water	108		
Eminent domain/easement	99		

Notes: An asterisk behind a word indicates coding included variations of the stem.

Critical discourse analysis (CDA) is both a method and a theory: "CDA is a helpful approach for scholars interested in examining mass media's roles in the social construction of reality" (Reynolds 2018, 48). Media analysis uses two different terms for the themes emerging from such analysis: "framing (or frame) analysis" in mass communication research and "holistic codes" used in CDA, with both being part of a single analytical frameworks "family" (Reynolds 2018, 59). All media contains frames "used strategically to frame a particular topic—like a picture frame that accentuates certain things, hides others, and borders off reality in a certain way . . . frame analysis looks at how existing 'objects' or 'topics' are framed by different actors, bending their meaning in certain directions" (Lindekilde 2014, 200). Coding is the process of analyzing the frames (and other information) to consider the ways in media constructs meaning. While there are similarities, each analysis addresses slightly different research questions. If the researcher is interested in how content is being presented to an audience, framing analysis is logical. If the researcher is working with a larger data set and wishes to systematically break content into points or ideas to illuminate categories and themes, coding is used (Erlingsson and Brysiewicz 2017). For the purposes of this study, both were utilized: coding to tackle the breadth of the media coverage, but framing for presentation of the Plains.

Language reflects mind-set while it constructs reality: "our lived experiences (for example, our thoughts, feelings, motivations, and the meaning we attribute to the world 'out there' are actively created through day-to-day linguistic practices and that process is, in turn, delimited by the shared language that culture makes available to us" (Di Masso et al. 2014, 75). Place meanings, often assumed to be individualistic, can be reinforced through interactions, where "place meanings become collectively shared, disseminated, and deployed" (Di Masso et al. 2014, 79). By publicly expressing place attachment, "we may be doing far more than expressing internal feelings that would otherwise remain within the private recesses of our minds . . . we may be performing social actions whose situated effects, quite literally, can move mountains" (Di Masso et al. 2014, 81).

In addition to considering the conversations taking place about the Plains and the pipelines, attention was paid to who was doing the "place talk," what was being said about place, and if place attachment manifested in their quotations (Di Masso and Dixon 2015).

Results: Themes, Place/Placelessness, and Place Attachment

Major Themes

With the media coverage of the KXL, the most frequently occurring themes were arguments in favor of the pipeline's construction (creating jobs, improved safety, energy security, less dependence on foreign oil). After this, environmental themes dominate: the Ogallala Aquifer, mentions of oil spills, the Sandhills, and concerns over the water supply and/or drinking water. The greater usage of "tar sands" suggests stronger proenvironmental framing, as proenergy and business framings use "oil sands" (Eilperin and Mufson 2013). Language of conflict is significant, but not as extensive as with the DAPL, reflecting fewer protests to date. Eminent domain/easement was a significant theme, as expected. This was one of the major issues, along with the Ogallala Aquifer. The Sandhills theme, while separate, is part of the Ogallala argument: a spill in the Sandhills could contaminate the underlying aquifer, a critical source of water for drinking and agriculture.

With DAPL, the media coverage overwhelmingly focused on conflict. The protests against DAPL were strong and the language used by the media emphasizes the conflict, often playing on the contentious history of Plains tribes and the U.S. government (such as "last stand," or "taking a stand," invoking "Custer's Last Stand").[6] DAPL's violent protests, extensively covered by the media, reflect the speed of pipeline construction and the protestors' determination to stop it, resulting in police and pipeline-security aggression.

Despite the focus on conflict, the prevalence of the sacred/spiritual theme ("sacred fire," "prayer," "spirit camp") reflects the Standing Rock leadership's framing of their actions (see Gass 2016a; 2016b). David Archambault II, then chairman of the Standing Rock Sioux, is quoted saying "We don't have weapons . . . We have people and prayer" (Worland 2016). "Water protectors" is what the resistance preferred to be called and invokes their belief in the sanctity of water (mni wiconi or "water is life") and the need to protect it for all (Enzinna 2017, 32).

There also occurred media discussions of rights, sovereignty, and treaties. The site of the pipeline's crossing of the Missouri was just outside of the Standing Rock Sioux Reservation but on lands granted to the Great Sioux Nation in the Fort Laramie Treaty of 1868.[7] DAPL resistance revived discussions of Native American rights and sovereignty and of the U.S. government's history of broken treaties.

Finally, as expected, in DAPL media coverage, there were recurrent themes of water supply/drinking water and of oil spills/leaks/breaks. There was significantly less discussion of the positive economic impacts of DAPL than with KXL.

PLACE/PLACELESSNESS

In presenting the "where" in news coverage, the media attempts to provide more than an absolute location, but pipelines present challenges when it comes to sense of place as they impact many locations. With the KXL media coverage, there is not one focal point; the potential construction site is thousands of miles long and protests largely diffused. Media coverage, as a result, tends to describe small town meetings or farmers or ranchers interviewed on their land.

When journalists did provide a sense of place, they tended to emphasize the rural landscape or attempted to capture a sense of the Plains:

> The wide open spaces of central Nebraska are an unlikely battlefield, but this is where the two opposing sides come head-to-head in the fight over the Keystone XL pipeline.
>
> Dean Reynolds (2014)

Or they try to convey the Sandhills:

> Gotschall and Kleeb were concerned that Keystone XL would cross the Sandhills, a 12 million-acre landscape of fragile, erodible soil and rolling sand dunes. A patch of trampled grass there can weather into a dune blowout and destroy a hay meadow.
>
> Madeline Ostrander (2012)

One of the few locations to receive repeated attention was Art and Helen Tanderup's farm:

> From the edge of rye field teeming with grasshoppers, Willie Nelson and Neil Young sang on Saturday in opposition to the proposed Keystone XL project . . . The site of the concert—a patch of farmland where 26 acres of corn were harvested early to create a makeshift parking lot . . . Their land in the rolling hills of northeastern Nebraska would be directly along the pipeline route.
>
> Mitch Smith (2014)

As the site of the 2014 "Harvest of Hope" concert, of an annual crop art instillation, and of an annual planting of Ponca sacred corn, on land donated to the tribe, the location (and the Tanderups) have become one of the go-tos for the media (Fahy 2017; Wernick 2017). As evident from the examples above, a sense of place for specific Plains locations is not provided, rather it is a generic space, using phrasing like "wide open," "rolling hills," and references to rural landscapes ("hay meadow," "rye field").

In contrast, the DAPL protests were more situated, taking place along the advancing construction sites leading to the Missouri River crossing:

> Horseback riders, their faces streaked in yellow and black paint, led the procession out of their tepee-dotted camp. Two hundred people followed, making their daily walk a mile up a rural highway to a patch of prairie grass and excavated dirt that has become a new kind of battlefield, between a pipeline and American Indians who say it will threaten water supplies and sacred lands.
>
> Jack Healy (2016a)

> The Standing Rock Sioux reservation sits in the Dakota Prairie Grasslands, an endless sweep of elephantine hills once home to millions of members of the Lakota Nation. Today, it's inhabited by few than 9,000 of their surviving descendants, and one of the few places in America where buffalo roam wild . . . .Their encampments of teepees, tents, and RVs were mostly ignored by the media until private security guards set dogs on protesters and a few journalists were arrested, sparking a national conversation about tribal sovereignty, environmental racism, and police brutality.
>
> Wes Enzinna (2017)

These descriptions of the protest camps are of a very specific location, yet they are also generic, with "rural," "prairie," and even buffalo roaming. The camp descriptions almost seem to describe stereotypical Native American existence 150 years ago, with references to horseback, teepees, face paint, and battles.

In both cases, there are challenges in presenting a sense of place; it is not a single place or town that will be impacted by the pipelines, but a 1,600-mile corridor of places. For KXL, there is an attempt to capture the Plains landscape, but journalists produce generic plains landscapes, reinforcing a sense of placelessness. With DAPL, the construction sites, protests, and camps received the most coverage, yet the descriptions conveying both a sense of placelessness and a sense of timelessness. The Plains, much as they were 150 years ago, are constructed as a plain space to be utilized for economic gain.

PLACE ATTACHMENT

In the KXL and DAPL coverage, a range of individuals are quoted on the pipelines. With KXL, those who express place attachment are primarily local residents, largely farmers and ranchers. With DAPL, those expressing place attachment are primarily Native American. Place attachment is a crosscutting theme in the media, manifesting throughout the coverage. In particular, place attachment is expressed in terms of threats: to home, to a valued resource, and to the future, tied broadly to Scannell and Gifford's functions that place attachment serves to humans (2010). Scannell and Gifford focus on the ways places serve humans: for survival and security, for goal support, for continuity, and for identity and belonging. The threat to home is a threat to survival and security, to continuity, as well as to identity and belonging. The threat to existence is a threat to survival and security, to continuity, and to belonging. The threat to the future is a threat to survival, to goal support, and to continuity.

Threats to home

Home for some is a refuge; for others, it's a prison. For those with a positive home experience and deep attachment, a threat to home can be profound, shaking the foundations of identity (Brown and Perkins 1992; Nowicki 2014). Threats to home can mobilize populations, near and far, to protect locations they call home (Carrus et al. 2014) as expressed by Nancy Zorn: "I can no longer sit by idly while toxic tar sands are pumped down from Canada and into our communities . . . It is time to rise up and defend our home" (quoted in Chamberlain 2013). "Home" manifested on a variety of levels, from the familial to the broader concept of homeland.

A number of individuals cite place attachment to land their families homesteaded or they currently farm: "'I don't want something like this to happen to what my grandparents built up,'" said Helen Tanderup (Struck 2017). With the Standing Rock community, it is a more complicated landscape:

> Verna Bailey stared into the silvery ripples of a man-made lake, looking for the spot where she had been born. 'Out there,' she said pointing to the water. 'I lived down there with my grandmother and grandfather. We had a community there. Now it's all gone.'
>
> Jack Healy (2016b)

The heart of the reservation was inundated when a dam was constructed in the 1950s (Estes 2019, 133–5). Yet, despite the loss, they still associate it with home, as Bailey expresses.

This deep familial attachment has been drawn upon by those organizing resistance:

> 'We still have a bunch of family farmers on the land that their ancestors homesteaded,' said Jane Kleeb, chair of the Nebraska Democratic Party and long-time organizer of opposition to the Keystone XL. 'They have a deep emotional and cultural tie to the land and feel a responsibility that they must protect it.'
>
> Steven Mufson (2017).

Familial place ties may link an individual to parents and grandparents who lived on and worked the land, and can be a positive element of their identity.

A more profound argument is made with DAPL in the ancestral ties of the Great Sioux Nation to the landscape: home on a different scale.[10] The Nation was granted all of South Dakota west of the Missouri under the Fort Laramie Treaty, but the treaty was broken (Lennard 2017). Place attachment is expressed on both an individual and a broader community/tribal basis. Laurie Running Hawk is quoted saying, "'I'm here,' . . . 'You're not going to kick me out. This is my land'" (Healy 2016d). Darrell Seki stated, "'These lands that you call the United States, these were tribal lands. You guys are foreigners. You guys are trespassing'"

(Hughlett 2016). The place attachment may be familial, to a landscape that their ancestors have called home for generations, as well as having tribal ties, forming an important component of identity. The dispossession of the Nation creates a population with strong place attachment with few having legal land ownership.

Not ironically, a recurring frame for KXL resistance is of a foreign corporation (TransCanada) threatening Americans' lands. Many individuals were angry the United States and state governments supported a foreign company's plans in the name of economic development and its use of eminent domain. "'Does a foreign company have the right to take a Texan's land if they don't want to give it up?'" asked Julia Trigg Crawford (Mufson 2012). There are strong sentiments of insiders feeling their homes threatened by outsiders, of the possibility of domicide. Home threats can result in action being taken to "defend our homes," as Nancy Zorn articulates.

In fact, while the DAPL protest camps were not home to anyone initially, they became home through lived and shared experiences. Retha Henderson said of the Sacred Stone Camp, "'This is my home now'" (Healy 2016c).

Threats to existence

On the Plains, water is a valuable commodity seldom taken for granted. The Plains' variable rainfall leads to feast and famine, bumper crops some years and devastating drought at others. An individual's perception of water is tied to their cultural, social, political, and economic context, with individuals dependent on water for their livelihood, such as farmers and ranchers, having a markedly different perception than someone living in a different environment with abundant resources (Shriver and Peaden 2009, 145).

Plains citizens' concern over water come through clearly in the media, such as Art Tanderup's, "'Everything we do here is out of the aquifer ... We drink it, the livestock drink it, we irrigate with it, and we water our gardens with it. It's our water'" (Pierre-Louis 2017). But they also attempt to frame water protection pragmatically, considering its value to the greater region, as Greg Awtry says: "'You can't see the Oglala Aquifer, but it may be the single most valuable resource the country has ... '" (Reynolds 2014). Or Nick Tilson's "'The goal is to stop the pipeline, and it's not just for us ... We know there are 17 million people downstream from us. The problem is bad for whatever community is near this pipeline.'" (Heim 2016). They emphasize the implication of a spill beyond the immediate vicinity.

To threaten a water supply is to threaten survival, especially with the prospect of climate change, with Plains states likely facing a hotter, drier future (EPA 2016). Without the water provided by the aquifer and by the Missouri, will it be possible to continue to live and make a living on the Plains? This threatens continuity as well as a sense of belonging.

Plains agriculture has been struggling for decades. Plains tribes have been struggling for even longer. Drucilla Burns' lament—"'We're supposed to be the protectors of the land and water. My God, they took everything away from us. And now they want to take our water too?'"—captures a sentiment expressed across the Plains (Burns, quoted in Heim 2016).

Threats to a future on the plains

Threats to the future are tied to place functions of survival, goal support, and continuity. Future can be tied to identity with a particular group (such as a tribe or nation), as well as the goals of building a legacy that might be someday passed to children, grandchildren, and so on.

There is concern over the immediate future, of how will people survive if their crucial water supply is contaminated, as in Cyril Scott's quote: "'We need to start remembering that the earth is our mother and stop polluting her and start taking steps to preserve the land, water, and our grandchildren's future'" (Haq 2014). Scott's concerns are for the immediate future and for future generations. The consideration of the immediate future can also be seen in quotes used in the discussions of threats to home and to existence: without the water needed for both home and livelihood, how will they continue to survive? This is a threat to the place function of goal support. They are dependent on the water, to make a living and to continue to utilize the land. It is also tied to their lifestyle. Farming and ranching are not just a profession but also a lifestyle, one that may have a familial connection. Thus, it is also a threat to continuity.

The concern for future generations comes up quite often in the media, such as: "'You and me are probably lucky, because we won't see it … But our kids and our families and our descendants will have to deal with it one way or another'" (Pierre-Louis 2017). This concern over providing a future for their descendants is all about continuity, about the family line continuing, about a community continuing.

But this concern over water extends far beyond the protestor's descendants, to consider the future of the planet, as Shannon Graves expresses:

> 'When I first started this, it was about my house … I just wanted to protect my home.' That was six years ago. Now, Graves said, 'what was once just my home is now my world that I'm standing to protect.'
>
> Jillian Goodman (2017)

The sentiments express a long view, considering those who will call the Plains, as well as the planet, home, not only now but also in the future.

Discussion: Reflecting on "Who Loves This Land"

Chase Iron Eyes challenged his community to step up:

> 'Fighters, brothers and sisters. Come. Heed the call to defend this country against all enemies, foreign and domestic. We shall find out who loves this land, who is loyal to the water and who is a traitor to this land, to our water.'
>
> Michael Bennett Cohen (2017)

To the people who "love this land," are "loyal to the water," and whose voices are utilized by the press (such as Iron Eyes), they are defending their places: their homes, their family farms, the landscapes they make a living from, their ancestral lands, or those that they simply love. They are practicing place protection and some articulate their place attachment in their efforts to gain allies and block the pipelines. Native Americans and the descendants of settlers-colonizers relate to the Plains differently, given their diverse cultures and histories. Both are attached to this landscape for different reasons as constructed in the media (Williams and Stewart 1998, 22). Farmers and ranchers, largely descendants of settlers-colonists, are presented as attached to a familial home (family farm), the places where they grew up, the place where they live and/or work. Native Americans are presented as attached primarily to an ancestral land, to "a specific geography," though there are some tied to a familial home, too (Estes 2019, 70). Through media's use of quotations, farmers such as Jim Carlson, Art and Helen Tanderup, and members of the water protectors such as Chase Iron Eyes and Laurie Running Hawk, faces and voices are given to those who oppose the pipelines and who, in voicing their opposition, express their place attachment. Readers are clearly being shown "who is attached," who belongs to the Plains (Scannell and Gifford 2010, 2).

In addressing the question of "where" in their coverage of U.S. resistance to pipelines, the media attempts to construct a sense of place, but often creates an impression of an empty, uninhabited, or unused landscape. Words such as "rural" or "plain" suggest little to be attached to. Without a clear sense of the value of this land (societal, financial) that will be lost or impacted by the pipelines, there seems to be little reason not to put a pipeline in.

Some locations are clearly constructed, such as the camps and the Tanderup farm. These locations serve as nodes in the media coverage, providing "image-ability" to phenomena that are largely beyond the conception of the American public, inscribing these locations as sites of resistance, and essentially elevating these locations in the landscape, declaring " … themselves as places that in some way stand out from the surrounding area" (Relph 1976, 35). Moreover, they are distinct places in a landscape often perceived as "placeless."

Imageability also extends to the pipelines. Pipelines crisscross the world but, as many are buried, they are not apparent to most viewers of these landscapes. As David Bond points out, the current activism around pipeline protests is really "to trip up the tidy image of fossil fuel infrastructure as 'safe, silent, and unseen,' to borrow a popular industry description of pipelines … Oil pipelines have become, well, visible. And in that rising visibility, petro-networks are being opened up to new forms of accountability and refusal" (Bond 2015, 27). Pipelines are not only made visible, but they are placed and take place. By drawing attention to the placement of the pipelines, pipeline resisters are making visible these threats to place.

With the landscapes being crossed by the pipelines, we have the construction of both placelessness and place, both grounded in human experiences of place and place attachment. Relph wrote in 1976 that the media contributes to placelessness, providing "no roots, no sense of belonging to a place," yet with the pipeline media coverage we have placelessness and a sense of belonging. Liu and Freestone call for a reconsideration of this binary of place and placelessness:

> These attributes need not be mutually exclusive and a more effective paradigm responsive to the complexities and contradictions of locality, globality, culture, experience and subjectivity recognizes the possibilities of their simultaneity. Better understandings of both place and placelessness today start not with their polarity but their hybridity.
>
> Edgar Liu and Robert Freestone (2017)

In locations like the Plains, it is a place to those attached to it and placeless to those who are not—it is both, simultaneously. This hybridity, both placed and placeless, is at the core of the conflicts over the pipelines: being simultaneously a plane/plain to utilize for economic development and a place with a deeply attached population. And this hybridity clearly comes through the media.

The media, in covering the pipeline battles, are informing the public of events occurring in the country and attempting to provide the essential facts about the events (five h's and a w). This includes the "where"—where does this take place, where should it?—and "who"—who is affected, who will benefit, who will be harmed. But it also helps to explain "why" a conflict is happening over the pipelines. Yet by reading, identifying with, and perhaps sympathizing with "place talk," connections can be made to an imagined community that extend well beyond the Plains (Scannell and Gifford 2010, 4; Tonge et al. 2015, 732).

Toward Conclusions: "Our Tie to the Land"

Larry Wright Jr., like Jim Carlson, has ties to the land:

> 'When we talk about our tie to the land, it's not about ownership.' Wright says. 'It's the fact that our relatives, our ancestors are buried in these hills. And every year when the grass grows [and] things green up, their spirit, their DNA is in those trees, in those flowers, in the corn that's grown. That's who we are.'
>
> Justin Perkins (2018)

In telling the pipeline protest stories, the media utilizes "ties to the land" to explain why individuals, as well as groups, have ferociously fought, and continue to fight, the pipelines. These expressions reflect the important functions place attachment serves to the human species, in particular the attachment to home as a place of both memory as well as identity, the deep ties to both land and water needed for survival of both the species and the individual, and the concerns over the future.

Place still matters. Place attachment still matters. In our increasingly polarized landscape, polarized over politics and issues like climate change, tensions will continue as we attempt to balance projects seen in the "national interest" (for example, pipelines or dams) and the places and peoples impacted by such projects. The media has played and will continue to play a role in influencing public opinion of these projects. "Ties to the land" are a means for the media to explain why Plains people actively fought the KXL for the past 12 years and why the DAPL led to two intense years of clashes that became increasingly violent. The hybridity of place and placelessness, and the potential leveraging of place attachment, are important to consider in the face of continuing battles over development.

Notes

[1] "XL" stands for Export Limited. The southern Keystone XL leg from Cushing, Oklahoma to Texas refineries was completed in 2010.

[2] DAPL protest sites were far from cities and resources: camps provided food and shelter for the protesters.

[3] Plains-set articles are in contrast to those that are set in Washington, D.C. (protests, hearings), or that deal abstractly with financial aspects. "Straight news" is a plain account of news facts, without exaggeration. Feature articles are a compelling news narrative. Editorials and opinion pieces are written from a particular perspective, often persuasively, and were excluded.

[4] All the articles on the pipelines published during this period were not collected. This is an extensive sample. All media sources have their ideological standpoints and attempts were made to balance sources for a representative sample.

[5] The author was the sole coder and analyzer of the articles. The value of multiple coders to content analysis is acknowledged (Erlingsson and Brysiewicz 2017, 98). As the intent was a qualitative analysis, using the content analysis only to get a sense of the "big picture," having a single coder was viewed as acceptable.

[6] Also called the Battle of Little Big Horn or of Greasy Grass, the 1876 battle was a decisive victory for the tribes, defeating the 7th Cavalry Regiment of the U.S. Army and killing its leader, George Armstrong Custer (White 1991, 104 and 622-3).

[7] The Great Sioux Nation is composed of Seven Council Fires or tribes, subdivided into 18 bands, including the Hunkpapa of the Standing Rock Indian Reservation (Estes 2019, 69-70).

Orcid

Christina E. Dando http://orcid.org/0000-0003-4749-0388

References

Adger, W., J. Barnett, F. Chapin III, and H. Ellemor. 2011. This Must Be the Place: Underrepresentation of Identity and Meaning in Climate Change Decision-Making. *Global Environmental Politics* 11(2):1–25. doi:10.1162/GLEP_a_00051.

Allen, J. 1985. The Desert-Garden Continuum: Competing Views of the Great Plains in the Nineteenth Century. *Great Plains Quarterly* 5(4):207–220.

______. 1993. New World Encounters: Exploring the Great Plains of North America. *Great Plains Quarterly* 13(2):69–80.

Bennett Cohen, M. 2017. Anti-DAPL Activists Call for Return to Battle at Standing Rock. *Newsweek*, January 26.

Bond, D. 2015. The Promising Predicament of the Keystone XL Pipeline. *Anthropology Now* 7 (1):20–28. doi:10.1080/19492901.2015.11728303.

Brown, B., and D. Perkins. 1992. Disruptions in Place Attachment. In *Place Attachment*, edited by I. Altman and S. Low, 279–304. New York: Plenum Press

Budinksy, J., and S. Bryant. 2013. 'It's Not Easy Being Green': The Greenwashing of Environmental Discourses in Advertising. *Canadian Journal of Communication* 38(2):207–226.

Buttimer, A. 1980. Home, Reach, and the Sense of Place. In *The Human Experience of Space and Place*, edited by A. Buttimer and D. Seamon, 166–187. New York: St: Martin's Press

Carrus, G., M. Scopelliti, F. Fornara, M. Bonnes, and M. Bonaiuto. 2014. . Place Attachment, Community Identification, and Pro-Environmental Engagement. In *Place Attachment: Advances in Theory, Methods and Applications*, edited by L. Manzo and P. Devine-Wright, 154–164. London and New York: Routledge

Chamberlain, J. 2013. 'Time to Rise Up': Oklahoma Grandmother Bike-Locks Herself to KXL Pipeline Machinery. April 10. https://www.commondreams.org/news/2013/04/10/time-rise-oklahoma-grandmother-bike-locks-herself-kxl-pipeline-machinery.

Cheng, T., H. Wu, and L. Huang. 2013. The Influence of Place Attachment on the Relationship between Destination Attractiveness and Environmentally Responsible Behavior for Island Tourism in Penghu. *Taiwan. Journal of Sustainable Tourism* 21(8):1166–1187. doi:10.1080/09669582.2012.750329.

Dando, C. 2008. Constructing a Home on the Range: Homemaking in Early Twentieth Century Plains Photograph Albums. *Great Plains Quarterly* 28(2):105–133.

______. 2009. Deathscapes, Topocide, Domicide: The Plains in Contemporary Print Media. *Great Plains Quarterly* 29(2):95–119.

Davenport, C. 2021. The Keystone XL Pipeline Project Has Been Terminated. *New York Times*, June 9.

Devine-Wright, P. 2009. Rethinking NIMBYism: The Role of Place Attachment and Place Identity in Explaining Place-Protective Action. *Journal of Community & Applied Social Psychology* 19 (6):426–441. doi:10.1002/casp.1004.

______. 2012. Explaining 'NIMBY' Objections to a Power Line: The Role of Personal Place Attachment and Project-Related Factors. *Environment and Behavior* 45(6):761–781. doi:10.1177/0013916512440435.

______. 2013. Think Global, Act Local? The Relevance of Place Attachment and Place Identities in a Climate Changed World. *Global Environmental Change* 23(1):61–69. doi:10.1016/j.gloenvcha.2012.08.003.

Di Masso, A., J. Dixon, and K. Durrheim. 2014. Place Attachment as Discursive Practice. In *Place Attachment: Advances in Theory, Methods, and Applications*, edited by L. Manzo and P. Devine-Wright, 75–86. London and New York: Routledge

Di Masso, A., and J. Dixon. 2015. More than Words: Place, Discourse and the Struggle over Public Space in Barcelona. *Qualitative Research in Psychology* 12(1):45–60. doi:10.1080/14780887.2014.958387.

Eilperin, J., and S. Mufson. 2013. State Dept. Says Keystone XL Would Have Little Effect on Climate Change. *The Washington Post*, March 2.

Entman, R. 2007. Framing Bias: Media in the Distribution of Power. *Journal of Communication* 57 (1):163–173. doi:10.1111/j.1460-2466.2006.00336.x.

Environmental Protection Agency. 2016. What Climate Change Means for Nebraska. EPA 430-F-16-029. August.

Enzinna, W. 2017. Crude Awakening. *Mother Jones* 42:32–37.
Erlingsson, C., and P. Brysiewicz. 2017. A Hands-On Guide to Doing Content Analysis. *African Journal of Emergency Medicine* 7(3):93–99. doi:10.1016/j.afjem.2017.08.001.
Estes, N. 2019. *Our History Is the Future: Standing Rock versus the Dakota Access Pipeline, and the Long Tradition of Indigenous Resistance.* London and New York: Verso.
Fahy, J. 2017. Farmer Hero Update: Art Tanderup, Neligh, Nebraska. April 25. https://www.farmaid.org/blog/farmer-heroes/farmer-hero-update-art-tanderup-neligh-nebraska/.
Forliti, A. 2021. Buoyed by Keystone XL, Pipeline Opponents Want Biden to Act. *ABC News*, January 31. https://abcnews.go.com/US/wireStory/buoyed-keystone-xl-pipeline-opponents-biden-act-75595940.
Gass, H. 2016a. For Native Americans, Pipeline Sparks Climate Awakening. *The Christian Science Monitor*, November 17.
_____. 2016b. Behind Dakota Pipeline Protest: Native American Religious Revival. *The Christian Science Monitor*, November 1.
Gillies, R. 2021. Keystone XL Halted as Biden Revokes Permit. Associated Press, January 20. https://apnews.com/article/joe-biden-alberta-2fbcce48372f5c29c3ae6f6f93907a6d.
Goodman, J. 2017. Can Some Nebraska Farmers Kill the Keystone XL Pipeline? August 14. https://www.bloomberg.com/news/articles/2017-08-14/can-some-nebraska-farmers-kill-the-keystone-xl-pipeline.
Groves, C. 2015. The Bomb in My Backyard, the Serpent in My House: Environmental Justice, Risk, and the Colonisation of Attachment. *Environmental Politics* 24(6):853–873. doi:10.1080/09644016.2015.1067348.
Haq, H. 2014. Sioux Tribe Calls Keystone XL Approval 'Act of War'. *Christian Science Monitor*, November 18.
Healy, J. 2016a. Tension on the Plains as Tribes Move to Block a Pipeline. *New York Times*, August 24.
_____. 2016b. On the Frozen Dakota Prairie, Readying for One Final Stand. *New York Times*, December 4.
_____. 2016c. Tensions Rising over Pipeline, Tribes Dig In: 'We're Staying'. *New York Times*, October 11.
_____. 2016d. 'I Want to Win Someday': Tribes Make a Stand against Pipeline. *New York Times*, September 9.
Heim, J. 2016. Showdown over Oil Pipeline Becomes a National Movement for Native Americans. *The Washington Post*, September 7.
Hodges, H., and G. Stocking. 2016. A Pipeline of Tweets: Environmental Movements' Use of Twitter in Response to the Keystone XL Pipeline. *Environmental Politics* 25(2):223–247. doi:10.1080/09644016.2015.1105177.
Hughlett, M. 2016. Long, Complicated Road Led to Current Standoff at Standing Rock. *[Minneapolis] Star Tribune*, December 3.
Lennard, N. 2017. 6 Standing Rock Dakota Access Pipeline Protestors Face Unprecedented Federal Charges. *Esquire*, September 19.
Lindekilde, L. 2014. Discourse and Frame Analysis: In-Depth Analysis of Qualitative Data in Social Movement Research. In *Methodological Practices in Social Movement Research*, edited by D. Della Porta, 195–227. Oxford, U.K.: Oxford University Press.
Liu, E., and R. Freestone. 2017. Revisiting Place and Placelessness. In *Place and Placelessness Revisted*, edited by R. Freestone and E. Liu, 1–19. London and New York: Routledge
McMaken, R. 2016. The Real Reason to Oppose the Dakota Access Pipeline. November 29. https://mises.org/wire/real-reason-oppose-dakota-access-pipeline.
Meyer, R. 2017. Oil Is Flowing through the Dakota Access Pipeline. *The Atlantic*, June 9.
Miller, S. 2017. Dakota Access, Keystone XL: 2 Pipelines, 2 Flashpoints. *USA Today*, January 24.
Mufson, S. 2012 . Whose Land Is It? *The Washington Post*, July 29.
_____. 2017. Trump Promised to Build the Keystone XL. Three Votes in Nebraska Could Stop It. *The Washington Post*, August 8.
Nowicki, M. 2014. Rethinking Domicide: Towards an Expanded Critical Geography of Home. *Geography Compass* 8(11):785–795. doi:10.1111/gec3.12178.

Olstad, T. 2013. Personal Experiences and Public Place Creation. In *Place-Based Conservation: Perspectives from the Social Sciences*, edited by W. Stewart, D. Williams, and L. Krugar, 99–107. Berlin: Springer

Ostrander, M. 2012. Transpartisan Politics on the Plains. *The Nation*, February 20.

Perkins, J. 2018. A New Native-led Strategy for Fighting Keystone XL. *In These Times*, July 18.

Pierre-Louis, K. 2017. This Land Is (Still) Their Land. Meet the Nebraska Farmers Fighting Keystone XL. *Popular Science*, September 15.

Porteous, J. 1989. *Planned to Death: The Annihilation of a Place Called Howdendyke*. Toronto: University of Toronto Press.

Porteous, J., and S. Smith. 2001. *Domicide: The Global Destruction of Home*. Montreal: McGill-Queens University Press.

Relph, E. 1976. *Place and Placelessness*. London: Pion Limited.

Reynolds, C. 2018. Building Theory from Media Ideology: Coding for Power in Journalistic Discourse. *Journal of Communication Inquiry* 43(1):47–69. doi:10.1177/0196859918774797.

Reynolds, D. 2014. Nebraska Becomes Battlefield in Fight over Keystone XL Pipeline. Cbsnews. May 20. https://www.cbsnews.com/news/nebraska-becomes-battlefield-in-fight-over-keystone-pipeline/.

Scannell, L., and R. Gifford. 2010. Defining Place Attachment: A Tripartite Organizing Framework. *Journal of Environmental Psychology* 30(1):1–10. doi:10.1016/j.jenvp.2009.09.006.

_____, and _____. 2013. Personally Relevant Climate Change: The Role of Place Attachment and Local versus Global Message Framing in Engagement. *Environment and Behavior* 45(1):60–85. doi:10.1177/0013916511421196.

Schulte, G. 2011. Pipeline Foes Pin Hopes on Nebraska. *The Washington Post*, September 26.

Shriver, T., and C. Peaden. 2009. Frame Disputes in a Natural Resource Controversy: The Case of the Arbuckle Simpson Aquifer in South-Central Oklahoma. *Society and Natural Resources* 22 (2):143–157. doi:10.1080/08941920801973789.

Smith, J. 2017a. Putting Place Back in Place Attachment Research. In *Explorations in Place Attachment*, edited by J. Smith, 1–16. London: Routledge

Smith, J., and J. McAlister. 2015. Understanding Place Attachment to the County in the American Great Plains. *Geographical Review* 105(2):178–198. doi:10.1111/j.1931-0846.2014.12066.x.

Smith, M. 2014. Nebraskans Raise Voices in Keystone Pipeline Debate. *New York Times*, September 30.

_____. 2017b. Risen from the Grave, Keystone XL Pipeline Again Divides Nebraska. *New York Times*, April 17.

Struck, D. 2017. In the Heartland, a David and Goliath Battle over a Pipeline. *Christian Science Monitor*, August 17.

Sullivan, K. 2016. Voices from Standing Rock. *The Washington Post*, December 3.

Tonge, J., M. Ryan, S. Moore, and L. Beckley. 2015. The Effect of Place Attachment on Pro-environment Behavioral Intentions of Visitors to Coastal Natural Area Tourist Destinations. *Journal of Travel Research* 54(6):730–743. doi:10.1177/0047287514533010.

Tuan, Y.-F. 1980. Rootedness versus Sense of Place. *Landscape* 24(1):3–8.

Upham, P., K. Johansen, P. Bögel, S. Axon, J. Garard, and S. Carney. 2018. Harnessing Place Attachment for Local Climate Mitigation? Hypothesizing Connections between Broadening Representations of Place and Readiness for Change. *Local Environment* 23(9):912–919. doi:10.1080/13549839.2018.1488824.

Wernick, A. 2017. The Keystone XL Pipeline Fight Continues. PRI Living on Earth. May 27. https://www.pri.org/stories/2017-05-27/keystone-xl-pipeline-fight-continues.

White, R. 1991. *'It's Your Misfortune and None of My Own': A New History of the American West*. Norman: University of Oklahoma Press.

Williams, D., and S. Stewart. 1998. Sense of Place: An Elusive Concept that Is Finding a Home in Ecosystem Management. *Journal of Forestry* 96(5):18–23.

Worland, J. 2016. A High-Plains Showdown over the Dakota Access Pipeline. *Time* 188:12.

INDIGENEITY, DISPLACEMENT, AND REGIONAL PLACE ATTACHMENT AMONG IDPS FROM CRIMEA

AUSTIN CHARRON

ABSTRACT. Studies of indigenous place attachments tend to focus on the alternative environmental ontologies that animate and inform indigenous peoples' relationships with place, but rarely engage with the material and psychological impacts of colonialist practices—namely, displacement and dispossession of land. Russia's annexation of the Crimean Peninsula in 2014 triggered the internal displacement of tens of thousands of Crimeans to mainland Ukraine, including ethnic Russians, Ukrainians, and indigenous Crimean Tatars. This multiethnic flow of internally displaced peoples (IDPs) affords the opportunity to contrast the post-displacement attachments of both indigenous and non-indigenous peoples to the same place. Based on ethnographic fieldwork, this paper argues that Slavs' attachments to Crimea are principally individualistic and rooted in a reverence for the peninsula's natural environment, while Crimean Tatars possess an intergenerational narrative of national belonging to Crimea informed by cyclical patterns and experiences of colonialist displacement from their homeland. *Keywords: Crimea, indigeneity, internal displacement, Ukraine.*

Judging by long-term trends in place-attachment research, the most salient places to which we develop attachments tend to be rather small and intimate in scale; as Maria Lewicka (2011) observes, studies of place attachment focus most frequently on the scale of the home, neighborhood, or city. However, Lewicka also notes that strong place attachments do develop at the regional scale in regions that, for historical reasons, "have acquired strong nationalistic meaning" (212), while others have similarly affirmed that regions may serve as salient sites for the development of place attachment (Cuba and Hummon 1993; Laczko 2005; Shamai, Arnon, and Schnell 2012). Rarely cited among nationally significant regions with a strong sense of identity, the contested Crimean Peninsula is precisely such a region, where well-developed regional identities are informed by competing narratives of the peninsula's underlying significance to its three primary ethno-national communities: Russians, Ukrainians, and indigenous Crimean Tatars (Charron 2016).

Recent events serve to highlight the deeply nationalistic underpinnings of Crimean regional identity, as rhetorical justifications for Russia's annexation of Crimea in 2014 have relied primarily on a political narrative portraying the region as innately Russian and totemic of Russian national identity (Teper 2016). But while many in Crimea embrace this essentialist narrative of Crimea's relationship to Russia, some tens of thousands of Crimeans have tacitly rejected it by relocating to the Ukrainian mainland as internally displaced peoples (IDPs). Uninhibited by the Kremlin's strict limitations on speech contradicting this prescribed narrative, Crimean IDPs articulate alternative perspectives on the affective meanings of Crimea influencing their regional identities—perspectives now colored by the experience of displacement.

Internal displacement tends to impact indigenous peoples or other culturally and economically marginalized communities disproportionately to others (Birkeland 2009), but Crimean IDPs represent a broad cross-section of Crimea's population, consisting of ethnic Russians, Ukrainians, Crimean Tatars, and other small minorities marginalized politically by their rejection of Russia's occupation of Crimea. These unique circumstances present a rare opportunity to consider variations in place attachment among different ethnic communities with a shared experience of displacement from the same region. While the impacts of both sociocultural identities—including ethnicity and indigeneity—and displacement are frequent themes in place-attachment studies, they are seldomly considered in tandem.

Ethnic cleavages play a key role in how Crimean IDPs express attachment to Crimea following their displacement, with Crimean Tatar perspectives diverging significantly from those of Slavic Crimeans—i.e., both Ukrainians and Russians. These divergent perspectives are linked directly to the indigenous status of Crimean Tatars and the legacy of Russian/Soviet colonialism responsible for Crimea's Slavic majority today. But while most studies of indigenous place attachment focus primarily on the alternative ontologies of place that animate indigenous worldviews, little attention has been paid to how indigenous relationships with place are transformed through practices of colonialism—namely, displacement and dispossession of (home)land. Based on ethnographic fieldwork among Crimean IDPs, this paper contrasts Crimean Tatar and Slavic discourses of attachment to Crimea as viewed from a displaced vantage point. I argue that Slavic Crimeans' relationships with Crimea are principally individualistic and rooted in a reverence for Crimea's natural environment, while Crimean Tatars possess an intergenerational narrative of national belonging to Crimea developed during previous periods of displacement, which informs their attachment to Crimea following displaced to mainland Ukraine.

Sociocultural Differences and Displacement in Place Attachment

Places are constructed from complex and often divergent assemblages of meaning that mediate relationships between spaces and the peoples who inhabit them. Doreen Massey describes places as being "constructed out of a particular constellation of social relations, meeting and weaving together at a particular locus" (1994, 154), but also points to the multiplicitous and potentially contradictory meanings generated at these intersections (155). The conflicting but coexistent meanings imbuing a particular place are a function of culture; theorizing "landscapes" as the "symbolic environments created by human acts of conferring meaning to nature and the environment," Thomasa Gredier and Lorraine Garkovich argue that landscapes fundamentally reflect the values and beliefs of the cultural groups who construct and occupy them (1994, 1). Setha Low similarly argues that the affective bonds groups form with their physical and symbolic environments are extensions of their cultural identities (1992, 166). These works reflect a trend in place-attachment studies wherein sociocultural differences rather than environmental characteristics are emphasized as the primary factor differentiating discourses of place attachment. As Lewicka demonstrates, most published research on place attachment is focused on the "people" dimension of a "people-place-process" nexus at the core of place attachment itself (2011).

Several authors examine correlations between embodied sociocultural identities and the robustness of place perceptions, preferences, and attachments professed by those belonging to particular groups, with many advocating for greater consideration for how racial, ethnic, gendered, sexual, and class identities impact humans' "emotional relationships" with place (Manzo 2005). Many studies pertain to places of leisure and recreation, such as works by Randy Virden and Gordon Walker (1999) and Cassandra Johnson and J. Michael Bowker (2004) that investigate how race and collective memory influence attitudes toward wildlands in the United States. But as Kelley McClinchey (2016) urges, greater attention should be paid to the differences in place attachment among ethno-cultural groups in particular.

In this spirit, several place-attachment studies examine how indigenous and non-indigenous identities engender different relationships with place. As indigenous peoples often ascribe to cosmological, ecological, and metaphysical perspectives that conflict with the hegemonic worldviews espoused by colonizers, indigenous understandings of the relationships between people and their environments are often portrayed as more "spiritual" (Stokols 1990) or phenomenological (Casey 2001; Seamon 2014) than those of non-indigenous peoples. Soren Larsen and Jay Johnson argue that, while Western ontologies of place generally uphold a fundamental disjunction between self and environment, indigenous ontologies typically view place as "a holistic reality in which myriad human and non-human beings are interconnected via genealogies contained within a landscape that is … resonant of cosmological and social origins, events, and encounters" (2012, 10). Taking their argument further, Larsen and Johnson assert that, from indigenous perspectives, place itself is animated by its own inherent agency both independent and generative of the human consciousness embedded within place (2016, 151). It is this holistic understanding of humans' relationships with place that often informs studies contrasting indigenous place attachments with the more "instrumentalist" relationships (Stokols 1990) non-indigenous peoples experience with(in) place.

Several authors compare indigenous and non-indigenous relationships with place empirically, including research into contrasting perceptions of homelessness in Ontario (Kauppi, Pallard, and Shaikh 2013) and the Northwest Territories (J. Christensen 2013) in Canada, and senses of place among various hunter-gatherer societies located across the globe (Thompson 2016). Many empirical studies of indigenous place attachment also focus on alternative ontologies of place in New Zealand (Hay 1998; Carr 2004; Farminer 2013) and Australia (Moreton-Robinson 2003; Carter and others 2018). But while there is a sizeable literature concerning the ontologies that inform indigenous place attachments, rarely do these works engage with the material realities and subaltern positionalities of indigenous peoples that may also influence their relationships with place in colonial contexts. Namely, the effects of displacement from and dispossession of native land—virtually universal to the experiences of indigenous peoples in settler-colonialist societies (Coleman and others 2012)—is lacking in much of the indigenous place-attachment literature.

Several works do explore themes of mobility, displacement, loss, and their effects on place attachment, yet indigenous perspectives are typically absent here, too. Questions of (im)mobility are a recurring theme in place- attachment studies, representing one of the most frequently interrogated factors contributing to variations in attachment to places of residency (Lewicka 2011). Many authors affirm that strength of place attachment is positively correlated with length of residency in a particular place, especially at the scale of neighborhood or city (Hay 1998; Shamai and Ilatov 2005; Hernández and others 2007; Fleury-Bahi, Félonneau, and Marchand 2008; Gustafson 2009a). Others contend that high residential and/or leisurely mobility does not necessarily diminish place attachments, but often fosters multilocal, translocal, and multiscalar attachments instead (Gustafson 2001, 2009a, 2009b, 2014; Cheng 2005; Christensen and Jensen 2011). Some also explore disruptions to or loss of attachment without movement away from a given place, but rather through changes in the place itself (Brown and Perkins 1992; Albrecht 2005).

But as Erika Apfelbaum laments (2000, 1009), political, economic, and environmental upheavals have produced growing numbers of displaced peoples since the mid-twentieth century, and the often traumatic experience of displacement compels consideration for the impact of such "uprootedness" upon place attachments. Lewicka (2014) and Michael Strong (2018) argues that memory is an effective enabler of place attachment among displaced peoples—either to the place left behind or to the place of resettlement through engagement with the place's historical memory as a coping strategy for displacement. Focusing on the plight of marginalized peoples who have been serially displaced within urban contexts, Mindy Thompson Fullilove argues that such chronic precarity often erodes the bonds between people and place, but that attachments may nevertheless be preserved through the (re)imagining of a "place-that-was-and-might-be-again" (2014, 141–42).

Despite the staggering variety of cultures and idiosyncratic relationships with place that they represent, indigenous peoples the world over nevertheless share common experiences of displacement and/or dispossession from native land due to colonial practices of settlement, development, and resource extraction; yet the impacts of these experiences on indigenous place attachment remains conspicuously underresearched. While some studies do frame indigenous relationships to place as, in part, a product of colonization and displacement (Moreton-Robinson 2003; Christensen 2013), much of the relevant literature seeks cultural and/or ontological explanations for differences in how indigenous and non-indigenous peoples relate to place. This work is indeed vital, but we run the risk of cultural essentialism and perpetuating marginalization if we focus too singularly on ontology and ignore the realities of displacement that also shape indigenous relationships with place. For millions of indigenous peoples there is some understanding of a "place-that-was-and-might-be-again" (Fullilove 2014), whether it be the native land from which they have been removed and imagine returning to, or the land on which they remain yet watch as it transforms into something unfamiliar (Albrecht 2005). The case of Crimea, whose annexation in 2014 triggered waves of internal displacement to mainland Ukraine by both indigenous and non-indigenous peoples, offers a unique opportunity to interrogate the varied effects of displacement on place attachment for both groups. Indeed, Crimea is the site of a long historical struggle between indigenous peoples and colonizers that has resulted in cyclical patterns of displacement and dispossession.

Historical Background: Indigeneity and Colonialism in Crimea

The Crimean Peninsula has been a site of both conquest and refuge for millennia, where subsequent waves of invasion and settlement—by Tauris, Cimmerians, Scythians, Sarmatians, Goths, Huns, Khazars, Greeks, Romans, Turks, Armenians, Italians, Slavs, Mongols, Kipchaks, and Nogais, among others—created a diverse ethno-cultural palimpsest by the mid-thirteenth century. Split between the Ottoman Empire and the Crimean Khanate by 1475, Crimea was jointly ruled by two Islamic authorities, and, as Brian Glynn Williams (2001) argues, it was the influence of a singular religious culture under Islam that first stimulated the coalescence of Crimea's diverse peoples into a hybrid regional ethnos: the Crimean Tatars. Although both religion and language were the main catalysts that drove the synthesis of a unified Crimean Tatar ethnos, Williams (2001) contends that some sense of territorial cohesion and mutual belonging to Crimea likely stimulated the formation of a pan-Crimean regionalism as a central component of a nascent Crimean Tatar identity from an early period. Herein lies the crux of Crimean Tatar ethnic and territorially identity, along with their claims to indigeneity (Belitser 2017): descended from a unique blend of peoples and cultures infused together within the Crimean Peninsula (Vozgrin 1992; Voitovich 2009), the Crimean Tatars are a product of Crimea's history and geography, a people who could not have appeared anywhere else, and who therefore identify not only as indigenous to Crimea, but autochthonous to it.

Growing in power and eager to compete with rival European empires for territory and influence, the Russian Empire first invaded and annexed Crimea in 1783. For Russia, Crimea was a strategically valuable acquisition that allowed it to project its influence and strength deep into the Black Sea region, but it also had great symbolic value. Crimea represented Russian's first colonial incursion into the "Orient" according to contemporary European perceptions and Orientalist discourses within Russia itself (Schönle 2001), helping to establish Russia's imperial credentials alongside other European empires actively seeking colonial possessions in the wider region (Dickinson 2002). The popular myth of Russian Orthodoxy's inception in Crimea with the 988 baptism of Prince Vladimir of Kyivan Rus' was also used as a justification for annexing Crimea at a time when Russia was fashioning itself into the foremost center of Orthodox Christianity (Kozelsky 2010).

Crimea would continue to acquire symbolic meanings over subsequent centuries within Russia's imperial grasp. Its rugged beauty attracting some of Russia's greatest literary figures of the nineteenth century, Crimea inspired its own distinctive literary tradition within the larger Russian cannon, endearing it to readers throughout the empire (Sasse 2007). Crimea also became one of Russian's premier tourist and health resort destinations by the late-nineteenth century, further endearing the peninsula to the millions of Russian and Soviet citizens who would vacation there. Perhaps most importantly, Crimea withstood intensive sieges and assaults during both the Crimean War (1853-1856) and World War II, resulting in a profound mythology of national perseverance and militarism surrounding the peninsula and especially the twice-destroyed naval port of Sevastopol, transforming the city into a potent symbols of Russian/Soviet patriotism (Plokhy 2000; Qualls 2009). From this heady stew of national symbols and mythologies, a distinctly "Crimean-Russian" identity would eventually materialize and form the foundation of a regional place attachment among the millions of ethnic Russians who settled in Crimea beginning in the late-eighteenth century (Charron 2012, 2016).

But constructing a Russian and, later, Soviet mythology of Crimea necessitated the suppression and erasure of its indigenous culture. Beginning shortly after the 1783 annexation, Crimean lands were confiscated from their indigenous owners and gifted as estates to members of the Russian nobility, reducing many Crimean Tatars to rent-paying serfs or otherwise compelling them to emigrate (Kırımlı 1996). Crimea's early Russian colonizers further implemented repressive measures against the Crimean Tatars with the intent of driving them from the most valuable land; according to Crimean Tatar political leader Mustafa Dzhemilev, early Russian colonialist tactics in Crimea included,

> terrorism and systematic plundering of the civilian population; seizure of the most fertile lands by high Tsarist officials; the displacement of Crimean Tatars to territories unsuitable for farming, thus depriving them of their means of existence; and the harsh violation of the Crimean Tatars' religious beliefs. (Cemiloglu [Dzhemilev] 1995, 88)

These efforts precipitated the beginnings of a Crimean Tatar exodus from the peninsula, primarily to the Ottoman Empire where an embryonic Crimean Tatar diaspora was beginning to take shape. The trickle of Crimean Tatar emigration became a flood beginning in the 1850s with the Crimean War, when escalating imperial rivalries pitted the Russian Empire against the Ottoman, French, and British empires, with Crimea serving as the theater of war and site of militarized violence on a massive scale. Additional waves of Crimean Tatar emigrants arrived in Ottoman lands during the 1860s when news of Russia's ethnic cleansing of the Circassian people of the North Caucasus sparked fears that Crimean Tatars may be targeted next (Williams 2001).

With the peninsula's population depleted due to extensive emigration and the ravages of the Crimean War, Tsarist authorities began sponsoring the mass settlement of Slavic peasants in Crimea—primarily Russians and Ukrainians—to replenish the region's labor force (Starchenko 2013). Thus began a drastic shift in Crimea's ethnic demographics that would see Crimean Tatars reduced to roughly one-third of the regional population by 1897 (Vodarskii, Eliseeva, and Kabuzan 2003). Kemal Karpat (1985, 66) estimates that the Russian colonization of Crimea displaced at least 1.8 million Crimean Tatars between 1783 and 1922, and while this figure is disputed, Hakan Kırımlı (1996) maintains that the number of Crimean Tatars who left during the nineteenth century at least far exceeds the number that stayed. Those who remained in Crimea were subjected to extreme policies of Russification (Fisher 1978), further contributing to the dilution and loss of Crimean Tatar culture and language as Crimea was gradually remade into a Russian cultural space.

Comprising only about 25 percent of Crimea's population by the time the Soviet Union was formed in 1922, the Crimean Tatars then endured heavy-handed programs of *korenizatsiya*—meaning "root-making" or "indigenization"—meant to sanitize and codify "official" versions of their national culture and language in ways consistent with the socialist and secular ideologies of the state (Williams 2015). While the ultimate objective of korenizatsiya was to reduce national cultures to their cosmetic trappings in order to foster a robust pan-Soviet identity—to render them "national in form, socialist in content," as the credo went—Kaiser (1994) argues that these policies mostly had the opposite effect, increasing perceptions of national difference and raising national consciousness rooted within a prescribed, territorialized homeland. Crimean Tatar's national consciousness and their perception of Crimea as a homeland both predated the Soviet period, but it was early Soviet policies of korenizatsiya that molded Crimean Tatar national identity into its modern form, however flattened and diluted (Williams 2015).

On the night of May 18, 1944, just days after Crimea had been liberated from Nazi occupation and under direct orders from Joseph Stalin, all Crimean Tatars were forced from their homes and onto cattle cars bound for distant corners of the Soviet Union, but mostly Uzbekistan. The Crimean Tatar deportation was predicated on accusations of their collective collaboration with the Third Reich during its occupation of Crimea, which sources refute despite the accusation's continued promulgation to this day (Fisher 1978; Williams 2001; Pohl 2010). Some estimate that as much as 46 percent of the Crimean Tatar population (110,000 out of 238,500) died in transit during their deportation due to disease, starvation, dehydration, and trauma, and those who survived were officially deprived of their national identity; Soviet authorities banned the use of the Crimean Tatar language and removed Crimean Tatars from the list of recognized nationalities, expecting them to eventually assimilate into the Uzbek or other ethnic populations (Williams 1997, 238). Emptied of its indigenous people, Crimea was now populated almost entirely by Russians and Ukrainians, many of whom now occupied Crimean Tatars' former homes and lands. Mosques and cemeteries throughout the peninsula were destroyed while hundreds of toponyms were Russified, as "the [Soviet] totalitarian regime used its vast, Orwellian resources to obliterate more than five hundred years of Tatar presence in the Crimean Peninsula" (238).

But in spite of the Soviet regime's efforts to erase them from memory, the Crimean Tatars persevered through the remainder of the Soviet period while remaining forbidden from returning to their homeland—an injustice that spawned the Crimean Tatar National Movement. One of the most ambitiously defiant and persistent protest movements in the history of the Soviet Union, the avowedly nonviolent Crimean Tatar National Movement fought singularly through petition and protest for the right to return to Crimea, resulting in criminal charges and imprisonment for many of its leaders (Alexeyeva 1987). The Crimean Tatars' connection to Crimea became the single most salient and galvanizing component of their national identity during this period of exile, functioning as a bulwark against cultural assimilation. Through tight-knit social networks, the memory of Crimea was kept alive and passed to younger generations who had never seen it for themselves, creating an "intergenerational narrative" (Merrill and Fivush 2016) of homeland and trauma from its deprivation (Uehling 2004; Bekirova 2017).

Authorities finally allowed the Crimean Tatars to return to their homeland in 1989, just as the Soviet Union was beginning to fracture under the weight of ethno-national demands for self-determination. Crimea had been transferred from the Russian SFSR to the Ukrainian SSR in 1954, but remained Russified linguistically and by population, with ethnic Russians now the clear majority. Although they had won a hard-fought battle for the right to live in their homeland, the Crimean Tatars remained politically marginalized and frequently vilified by Slavic Crimeans threatened by their return (Bezverkha 2017). As a pro-Russian separatist movement threatened Crimea's stability during the early 1990s (Sasse 2007), most Crimean Tatars aligned with Kyiv and supported Ukrainian sovereignty over Crimea in order to gird against further Russian incursions—a marriage of political convenience that would blossom into a genuine Ukrainian civic identity for many Crimean Tatars over the following decades (Charron 2018, 2019).

Comprising around 12 percent of the region's population by 2001 (State Statistics Committee of Ukraine 2004), the Crimean Tatars had again become integral yet continually marginalized members of Crimean society by the 2010s, while Crimea itself had become gradually entrenched as a Ukrainian region despite its Russian majority and persistent pro-Russian political sentiments. The latent threat of Russian revanchism became real in February 2014; immediately following the conclusion of Ukraine's Euromaidan Revolution, covert Russian forces seized control of the Crimean government and forced a fraudulent referendum on the region's status, paving the way for formal annexation on March 18, 2014. Russia's second annexation of Crimea—effectively a recolonization—received enthusiastic support from many Slavic Crimeans who remained bitter over the peninsula's placement within Ukraine instead of the Russian Federation since the Soviet Union's demise. Yet many others—particularly Crimean Tatars—favored Ukraine and had supported the Euromaidan's revolutionary thrust away from Russia, and hence were alarmed by Moscow's forceful seizure of Crimea and the oppressive political regime it imposed there (Coynash and Charron 2019). Weighing financial, political, and practical considerations (Charron 2020), tens of thousands of Crimeans of all ethnic backgrounds fled Crimea in the wake of the annexation and resettled in mainland Ukraine as IDPs.

Contrasting Crimean Tatar and Slavic Crimean IDP Attachments to Crimea

Although far outnumbered by the roughly 1.5 million people internally displaced by the violent conflict in the Donbas region of eastern Ukraine, Crimean IDPs now constitute a sizeable community spread across mainland Ukraine. While no official figures are available regarding the number of IDPs from Crimea specifically, they have been estimated at up to 100,000 (Blair 2016). The following discussion draws from semistructured interviews and focus groups conducted with Crimean IDPs in either Russian or English between September 2015 and July 2017, primarily in the cities of Kyiv (55 interviews, three focus groups) and Lviv (26 interviews). Because the unavailability of data concerning the numbers and demographic characteristics of

Crimean IDPs made obtaining a representative sample of this population impossible, I instead sought a diversity of perspectives according to age, gender, ethnicity, occupation, and education using a variety of nonrandom sampling methods. Participants were recruited through snowball sampling, opportunistic sampling at events and gatherings, and targeted sampling of more prominent members of the Crimean IDP community. Participants were asked a variety of questions pertaining to their experiences of displacement and the roles of both Crimea and Ukraine in their perceptions of territorial belonging.

Strong attachments to Crimea, manifested as a pronounced sense of regional identity, have been documented and theorized since long before the Russian occupation and internal displacement crisis began in 2014 (Kisileva 1999; Korostelina 2008; Knott 2015). But the nature of this regional identity and the affective meaning of Crimea itself varies tremendously for its residents depending on whose "national lens" one views it through (Charron 2012, 2016), and these differences are elucidated more clearly through the experience of displacement. As descendants and benefactors of a Russian/Soviet colonialist project that displaced Crimea's indigenous peoples and imposed an imperialist counternarrative about the region's underlying meaning and significance, Slavic Crimeans never experienced displacement or alienation from Crimea on a wide scale before 2014. Crimean Tatars, on the other hand, tend to view the 2014 annexation through the lens of their own historical memory, revealing it as another iteration in a centuries-long pattern of colonization and displacement that has ultimately strengthened their connection to Crimea and continually reinforced the imperative to return—an imperative with no such parallel among Slavic Crimeans.

Most Crimean Tatars are acutely aware that 2014 marks Russia's second annexation of Crimea, and that their relocation to mainland Ukraine represents their people's third major displacement event following the waves of emigration in the nineteenth century, and the 1944 deportation. "Catherine the Great, Stalin, and now Putin. One and the same, nothing ever changes," one focus group participant told me in frustration while reflecting on the Crimean Tatars' historical fate (Crimean Tatar woman, 40s). "We know what it means to be with Russia," an interviewee explained with regard to the Crimean Tatars' opposition to the 2014 annexation;

> We know the first annexation of Crimea by Catherine [the Great] in the eighteenth century, when much of the Crimean Tatar intelligentsia left ... We know the period of deportation in 1944, when half our population died. Historical memory has a very strong impact. (Crimean Tatar man, 20s)

Another interviewee similarly laid the blame for the Crimean Tatars' subjugation squarely at the feet of their colonizers: "[i]n their blood and from their mothers' milk, Crimean Tatars already have an antipathy toward this country; Russia has brought a lot of trouble to our nation" (Crimean Tatar man, 40s).

Incidentally, several Crimean Tatar interviewees conceptually evoked "mother's milk" as an intergenerational vector of inherited attachment to Crimea. "All of us certainly, from our mothers' milk, we know that Crimea is our homeland," one Uzbekistan-born focus group participant mused, explaining that,

> Our children are like this, and we were like this. I always remember how my grandmother cried as she remembered the place from where she was deported ... We probably didn't fully understand this feeling until we arrived in Crimea ourselves, and especially now that we have come to be in this situation, we ourselves think about Crimea and we understand that we are the same [as our grandmothers]. (Crimean Tatar woman, 40s)

This narrative of inherited belonging to Crimea also appears to be instilled within younger generations born in Crimea itself. Having developed personal attachments to their homeland since childhood—a privilege denied to their parents and grandparents—post-Soviet generations of Crimean Tatars nevertheless couch their sense of territorial belonging within this broader intergenerational narrative. As one Crimea-born focus group participant stated,

> We consider our homeland to be our most native place, which is Crimea ... We consider this a real homeland, where we were born and where we lived. And our parents consider this their homeland because it was passed to them through their mothers' milk. A person born in any given location will not necessarily consider that place his homeland, but instead the place where his parents or ancestors come from. (Crimean Tatar man, 20s)

In the words of one Crimean Tatar interviewee, "it seems like we are genetically linked to Crimea" (Crimean Tatar woman, 40s).

By contrast, most Slavic Crimeans' ancestral roots in Crimea—if they have any at all—are rather superficial. While some may be able to trace their lineage to the earliest Slavic settlers who arrived in the eighteenth century, anecdotal impressions are that most of Crimea's ethnic Russians and Ukrainians today arrived much more recently. "For 50 years there was really something like an overpopulation in Crimea, different people came and settled there," remarked one Ukrainian interviewee who had moved to Crimea only as a university student, adding, "among our [Slavic] friends and the people we interacted with in Crimea, there probably weren't any true [*chistye*] Crimeans. You could maybe count them on one hand" (ethnic Ukrainian man, 40s). "There are very few people who are native Crimeans. What is a native Crimean from our point of view? This is when at least your grandmother lived in Crimea. There are practically no Crimeans like this," explained another interviewee (ethnic Russian man, 50s). Slavic Crimeans therefore lack the discourse of a deep ancestral rootedness to anchor their senses of territorial belonging, and thus they relate and express attachments to Crimea principally on an individualistic basis. As one ethnic Russian interviewee stated, "[Crimea] has a monopoly on my childhood memories. I could live in New York for 30 years, I could live in London for 20 years, or live out the rest of my years in Kyiv, but these cities will never have my childhood memories, they are all in Crimea" (ethnic Russian man, 30s).

In the majority of cases, Slavic Crimean IDPs describe their relationship with Crimea primarily in terms of its natural environment; with its inviting coastline and rugged mountains, Crimea has long attracted artists, vacationers, and outdoor enthusiasts of all stripes, and Crimeans themselves typically cite the peninsula's natural splendors among its most defining characteristic (Charron 2012). "We love Crimea because it's beautiful," one ethnic Ukrainian interviewee told me,

> we love that the forest comes right up to the sea, that the sea is warm and inviting (…) We know that on warm days in November you can run down to the sea and go swimming. So, we know Crimea and love it for the way it is … It's beyond a national label—it's ours, we live there, and we love it. (ethnic Ukrainian man, 40s)

For many Slavic Crimean IDPs, it is through the dissonant feeling of displacement from one environment to another that they are able to articulate their underlying sense of attachment to Crimea.

"First of all, the sea. This is my first association with Crimea," one ethnic Russian interviewee responded when asked to describe Crimea's meaning to her. As she mused during our interview in her new home of Lviv, "[t]he sea just always called to me, and the water gives me a calming feeling, it gives me strength. There's nothing like that here. It's different here. There are [the Carpathian] mountains, but I think that everybody who grew up near the sea feels like I do. Crimea is for me, first of all, nature, mountains, the sea, everything around" (ethnic Russian woman, 20s). In perhaps the most doleful expression of longing for Crimea's natural environment that I encountered, one ethnic Russian woman also living in Lviv attributed her yearning for Crimea to an environmentally determined, physiological dependence:

> We all hurt in the same way; it's the air, the sea, we remember how it smelled there in the springtime. But what can we do? The trout lives in one kind of water, and the carp in a different kind. We can change the water, but it won't work out; The carp can only live in muddy water, and the trout needs fresh water … . We are a different people. It's very difficult for southerners to live here. We've come to Lviv, to this beautiful place, but for southerners it's really tough to live in a different region. It's genetic, there's no way to fool it. (ethnic Russian woman, 50s)

On the surface, this appeal to genetics draws an intriguing parallel with the claim of Crimean Tatars' genetically encoded belonging to Crimea quoted above. But here the notion works only as a metaphor for environmental affinity, while Crimean Tatars mostly do descend from the uniquely Crimean crucible of peoples from which their ethnos and ethnic consciousness first materialized.

For their part, Crimean Tatar IDPs also express longing for Crimea's natural environment and for the places that hold personal meaning to them. In one example, a Crimean Tatar man now living in Lviv described his emotional connection to Crimea's natural environment in terms that resemble those offered by several Slavic Crimean IDPs:

> For me, Crimea, the salty sea water, the wind, it's like a way of life, a basic necessity. It was tough for me in Lviv at first, there wasn't enough sun, it was constantly rainy, and far from the sea. In the summertime you want to go swimming, but here there are only some puddles with too many people. There's no space, and there's no place where you can climb up high and have your breath taken away because you feel the power of nature below, along with all the aromas. There's nothing like that here. The Carpathians have their own charms, but the Crimean Mountains are different. I really miss Crimea so much. (Crimean Tatar man, 40s)

However, his elaborations reveal that this relationship to Crimea's natural environment is inextricable from the intergenerational narrative of national belonging and return that he inherited as a child in Uzbekistan, which colored his impressions of the Crimean homeland long before he ever set foot there:

> I was born in Uzbekistan, outside of Tashkent, and for 15 years I heard about Crimea, that there is this promised land somewhere, our homeland, where my great-grandfather and great-grandmother lived, and where my great-grandfather's house is. I heard all these stories, and I always knew that we would return, even in the 70s and 80s when I was still a little boy. We always talked about this, even when it was impossible, it was a necessity for us … I've been dreaming about [Crimea] lately. I dream about the White Cliffs, where I have been to the very top, where I explored all the caves. I dream about Demerdzhi Mountain as I approach the sea, I dream about everything connected with this. I dream about my trees that grow in my courtyard, the grape vines, and this is all intermixed with the fact that it's connected to my grandmother and grandfather. (Crimean Tatar man, 40s)

In contrast to their Slavic counterparts, Crimean Tatars' hard-fought claims of indigenous belonging to Crimea are so deeply encoded into their narrative of national identity that they transcend reverence for the appealing characteristics of the place itself. Mustafa Dzhemilev, esteemed veteran of the Crimean Tatar National Movement and the Crimean Tatar's preeminent political leader today, summarized their relationship to Crimea thusly:

> During the few decades after their deportation, the Crimean Tatars fought desperately to return to Crimea not because it's beautiful there, with the Black Sea and the mountains. We probably would have struggled to return even if it was just a barren steppe, because that is where our nation was formed, that is where the graves of our ancestors are located. Any further perspectives for the existence and development of our nation are found only in our own land. (Dzhemilev 2016)

It is this indefatigable connection and commitment to homeland that animates the Crimean Tatars' attachment to Crimea, and which has enabled them to endure colonialist practices of displacement, dispossession, and disruption with their territorially rooted national identity intact. The Russian annexation of Crimea in 2014 opened a new chapter in this cyclical pattern of colonialist violence, this time displacing many Slavic Crimeans along with Crimean

Tatars, but it is the condition of rooted indigeneity that fortifies the Crimean Tatars' attachment to Crimea in a way impossible for Slavs to replicate.

Conclusions

Crimea is fraught with contested meanings that variably undergird regional place attachments for both its Slavic and indigenous Crimean Tatar populations. The multiethnic flow of internally displaced peoples to mainland Ukraine following Russia's annexation of Crimea in 2014 offers an opportunity to contrast place attachments to Crimea both through the lens of displacement and according to differences in ethno-national identity. An ethnographic study of Crimean IDPs reveals that Slavs' attachments to Crimea are principally individualistic and tied to their reverence for Crimea's natural environment. As such, Slavic Crimean IDPs generally express loss and longing for Crimea in terms of the rupture to their emotional connections with Crimea's natural places, especially the sea and the mountains. But while Slavic Crimeans have never before experienced displacement on such a scale, the recent flow of displaced Crimean Tatars is just the latest iteration in a cyclical pattern of colonially driven displacement that began with the first annexation of Crimea in 1783 and continued with the 1944 Crimean Tatar deportation. That experience of displacement and exile resulted in an entrenched intergenerational narrative of national belonging to Crimea that helped inoculate the Crimean Tatars against assimilation and propelled their activistic demands for the right to live in their homeland. It is this same narrative of indigeneity and intergenerational belonging to Crimea that informs the identities and place attachments of the Crimean Tatars now displaced in mainland Ukraine. While much of the literature concerning indigenous place attachment crucially emphasizes the non-Western ontologies that inform indigenous perceptions of place, the case of Crimean IDPs helps demonstrate that greater attention should also be paid to the impacts of colonialist policies and practices on indigenous peoples' relationships to place.

Funding

This work was supported by the National Science Foundation, Division of Arctic Sciences [1546954]. This article was prepared in part while the author held a postdoctoral fellowship at the University of Wisconsin-Madison with the Wisconsin Russia Project, sponsored by Carnegie Corporation of New York.

References

Albrecht, G. 2005. "Solastalgia." A New Concept in Health and Identity. *PAN: Philosophy Activism Nature* 3: 41–55.

Alexeyeva, L. 1987. *Soviet Dissent: Contemporary Movements for National, Religious, and Human Rights*. Middletown, Conn: Wesleyan University Press.

Apfelbaum, E. R. 2000. And Now What, After Such Tribulations? Memory and Dislocation in the Era of Uprooting. *American Psychologist* 55 (9): 1008–13.

Bekirova, G. 2017. *Piv Stolittia Oporu: Kryms'ki Tatary Vid Vyhnannia Do Povernennia (1941–1991 Roky)*. Kyiv, Ukraine: Krytyka.
Belitser, N. 2017. *Kryms'ki Tatary Iak Korinnyi Narod*. Kyiv, Ukraine: Natsional'ne gazetno-zhurnal'ne vidavnytsvo.
Bezverkha, A. 2017. Reinstating Social Borders between the Slavic Majority and the Tatar Population of Crimea: Media Representation of the Contested Memory of the Crimean Tatars' Deportation. *Journal of Borderlands Studies* 32 (2): 127–139.
Birkeland, N. M. 2009. Internal Displacement: Global Trends in Conflict-Induced Displacement. *International Review of the Red Cross* 91 (875): 491–508.
Blair, D. 2016. 100,000 Flee 'Worsening Oppression' as Russia Tightens Grip on Crimea. *The Telegraph*, 30 November. [http://www.telegraph.co.uk/news/2016/06/11/100000-flee-crimea-as-russia-tightens-grip/].
Brown, B. B., and D. D. Perkins. 1992. Disruptions in Place Attachment. In *Place Attachment*, edited by I. Altman and S. M. Low, 279–304. New York: Plenum Press.
Carr, A. 2004. Mountain Places, Cultural Spaces: The Interpretation of Culturally Significant Landscapes. *Journal of Sustainable Tourism* 12 (5): 432–459.
Carter, J., D. Hollinsworth, M. Raciti, and K. Gilbey. 2018. Academic "Place-Making": Fostering Attachment, Belonging and Identity for Indigenous Students in Australian Universities. *Teaching in Higher Education* 23 (2): 243–260.
Casey, E. S. 2001. Between Geography and Philosophy: What Does It Mean to Be in the Place-World? *Annals of the Association of American Geographers* 91 (4): 683–93.
Cemiloglu [Dzhemilev], M. 1995. A History of the Crimean Tatar National Liberation Movement: A Sociopolitical Perspective. In *Crimea: Dynamics, Challenges, and Prospects*, edited by Maria Drohobycky, 87–105. Lanham, Md.: Rowman & Littlefield.
Charron, A. 2012. "Through the National Lens": Nationality, Territory, and the Formation of 'Crimean-Russian' Identity. MA thesis, University of Kansas.
_____ 2016. Whose Is Crimea? Contested Sovereignty and Regional Identity. *Region: Regional Studies of Russia, Eastern Europe, and Central Asia* 5 (2): 225–256.
_____ 2018. In Our Country, but Outside Our Homeland: Identity and Diaspora Among Ukraine's Internally Displaced Crimeans. PhD diss., University of Kansas.
_____ 2019. Crimean Tatars' Postcolonial Condition and Strategies of Cultural Decolonization in Mainland Ukraine. *Euxeinos: Governance and Culture in the Black Sea Region* 9 (28): 26–46.
_____ 2020. "Somehow, We Cannot Accept It": Drivers of Internal Displacement from Crimea and the Forced/Voluntary Migration Binary. *Europe-Asia Studies* 72 (3): 432–454.
Cheng, H. L. 2005. Constructing a Transnational, Multilocal Sense of Belonging: An Analysis of *Ming Pao (West Canadian Edition)*. *Journal of Communication Inquiry* 29 (2): 141–159.
Christensen, A-D., and S. Q. Jensen. 2011. Roots and Routes. *Nordic Journal of Migration Research* 1 (3): 146–55.
Christensen, J. 2013. "Our Home, Our Way of Life": Spiritual Homelessness and the Sociocultural Dimensions of Indigenous Homelessness in the Northwest Territories (NWT), Canada. *Social & Cultural Geography* 14 (7): 804–828.
Coleman, D., E. G. Glanville, W. Hasan, and A. Kramer-Hamstra, eds. 2012. *Countering Displacement: The Creativity and Resilience of Indigenous and Refugee-ed Peoples*. Edmonton, Canada: University of Alberta Press.
Coynash, H., and A. Charron. 2019. Russian-Occupied Crimea and the State of Exception: Repression, Persecution, and Human Rights Violations. *Eurasian Geography and Economics* 60 (1): 28–53.
Cuba, L., and D. M. Hummon. 1993. A Place to Call Home: Identification with Dwelling, Community, and Region. *The Sociological Quarterly* 34 (1): 111–131.
Dickinson, S. 2002. Russia's First "Orient": Characterizing the Crimea in 1787. *Kritika: Explorations in Russian and Eurasian History* 3 (1): 3–25.
Dzhemilev, M. 2016. Interview with the author. Kyiv, Ukraine, 6 February.
Farminer, A. S. 2013. People, Rivers and Recreation: Fluid Relationships of Place and Experience on the Clutha River, Otago, New Zealand. PhD diss., University of Otago.
Fisher, A. W. 1978. *The Crimean Tatars*. Stanford, Calif.: Hoover Institution Press.

Fleury-Bahi, G., M-L. Félonneau, and D. Marchand. 2008. Processes of Place Identification and Residential Satisfaction. *Environment and Behavior* 40 (5): 669–682.
Fullilove, M. T. 2014. "The Frayed Knot": What Happens to Place Attachment in the Context of Serial Forced Displacement? In *Place Attachment: Advances in Theory*, Methods *and* Applications, edited by. L. C. Manzo and P. Devine-Wright, 141–153. New York: Routledge.
Greider, T., and L. Garkovich. 1994. Landscapes: The Social Construction of Nature and the Nvironment. *Rural Sociology* 59 (1): 1–24.
Gustafson, P. 2001. Roots and Routes: Exploring the Relationship Between Place Attachment and Mobility. *Environment and Behavior* 33 (5): 667–686.
_____ 2009a. Mobility and Territorial Belonging. *Environment and Behavior* 41 (4): 490–508.
_____ 2009b. More Cosmopolitan, No Less Local: The Orientations of International Travellers. *European Societies* 11 (1): 25–47.
_____ 2014. Place Attachment in an Age of Mobility. In *Place Attachment: Advances in Theory, Methods and Applications*, edited by L. C. Manzo and P. Devine-Wright, 37–48. New York: Routledge.
Hay, R. 1998. A Rooted Sense of Place in Cross-Cultural Perspective. *Canadian Geographer* 42 (3): 245–266.
Hernández, B., M. C. Hidalgo, M. E. Salazar-Laplace, and S. Hess. 2007. Place Attachment and Place Identity in Natives and Non-Natives. *Journal of Environmental Psychology* 27 (4): 310–319.
Johnson, C. Y., and J. M. Bowker. 2004. African-American Wildland Memories. *Environmental Ethics* 26 (1): 57–75.
Kaiser, R. 1994. *The Geography of Nationalism in Russia and the USSR*. Princeton, N.J.: Princeton University Press.
Karpat, K. H. 1985. *Ottoman Population 1830-1914: Demographic and Social Characteristics*. Madison: University of Wisconsin Press.
Kauppi, C., H. Pallard, and A. Shaikh. 2013. Migration and Homelessness: Exploring Attachment to Place amongst Francophone, Anglophone and Indigenous People in Northeastern Ontario. *Spaces & Flows: An International Journal of Urban & Extra Urban Studies* 3 (1): 97–106.
Kisileva, N. 1999. Za Perekopom Dlia Nas Zemli Net! *Ostrov Krym* 3: 44–45.
Kırımlı, H. 1996. *National Movements and National Identity among the Crimean Tatars (1905–1916)*. New York: Brill.
Knott, E. 2015. What Does It Mean to Be a Kin Majority? Analyzing Romanian Identity in Moldova and Russian Identity in Crimea from Below. *Social Science Quarterly* 96 (3): 830–59.
Korostelina, K. 2008. Identity, Autonomy, and Conflict in Republics of Russia and Ukraine. *Communist and Post-Communist Studies* 41 (1): 79–91.
Kozelsky, M. 2010. *Christianizing Crimea: Shaping Sacred Space in the Russian Empire and Beyond*. DeKalb: Northern Illinois University Press.
Laczko, L. S. 2005. National and Local Attachments in a Changing World System: Evidence from an International Survey. *International Review of Sociology* 15 (3): 517–528.
Larsen, S. C., and J. T. Johnson. 2012. In Between Worlds: Place, Experience, and Research in Indigenous Geography. *Journal of Cultural Geography* 29 (1): 1–13.
_____, and _____. 2016. The Agency of Place: Toward a More-Than-Human Geographical Self. *GeoHumanities* 2 (1): 149–166.
Lewicka, M. 2011. Place Attachment: How Far Have We Come in the Last 40 Years? *Journal of Environmental Psychology* 31 (3): 207–230.
_____ 2014. In Search of Roots: Memory as Enabler of Place Attachment. In *Place Attachment: Advances in Theory, Methods and Applications*, edited by L. C. Manzo and P. Devine-Wright, 49–60. New York: Routledge.
Low, S. M. 1992. Symbolic Ties that Bind. In *Place Attachment*, edited by I. Altman and S. M. Low, 165–185. New York: Plenum Press.
Manzo, L. C. 2005. For Better or Worse: Exploring Multiple Dimensions of Place Meaning. *Journal of Environmental Psychology* 25 (1): 67–86.
Massey, D. 1994. *Space, Place, and Gender*. Minneapolis: University of Minnesota Press.

McClinchey, K. A. 2016. Conceptualizing Sense of Place through Multi-Ethnic Narratives at a Multicultural Festival. Travel and Tourism Research Association: Advancing Tourism Research Globally 2. [https://scholarworks.umass.edu/ttra/2011/Student/2].

Merrill, N., and R. Fivush. 2016. Intergenerational Narratives and Identity Across Development. *Developmental Review* 40: 72–92.

Moreton-Robinson, A. 2003. I Still Call Australia Home: Indigenous Belonging and Place in a White Postcolonizing Society. In *Uprootings/Regroundings: Questions of Home and Migration*, edited by S. Ahmed, C. Castañeda, A-M. Fortier, and M. Sheller, 23–40. Oxford, U.K.: Berg.

Plokhy, S. 2000. The City of Glory: Sevastopol in Russian Historical Mythology. *Journal of Contemporary History* 35 (3): 369–383.

Pohl, J. O. 2010. The False Charges of Treason against the Crimean Tatars. International Committee for Crimea. [http://www.iccrimea.org/scholarly/pohl20100518.pdf].

Qualls, K. D. 2009. *From Ruins to Reconstruction: Urban Identity in Soviet Sevastopol after World War II.* Ithaca, N.Y.: Cornell University Press.

Sasse, G. 2007. *The Crimea Question: Identity, Transition, and Conflict.* Cambridge, Mass.: Harvard University Press.

Schönle, A. 2001. Garden of the Empire: Catherine's Appropriation of the Crimea. *Slavic Review* 60 (1): 1–23.

Seamon, D. 2014. Place Attachment and Phenomenology: The Synergistic Dynamism of Place. In *Place Attachment: Advances in Theory, Methods and Applications*, edited by L. C. Manzo and P. Devine-Wright, 11–22. New York: Routledge.

Shamai, S., S. Arnon, and I. Schnell. 2012. From Home to Community and Settlement: Sense of Place in Different Scales. *Studies on Home and Community Science* 6 (3): 153–163.

Shamai, S., and Z. Ilatov. 2005. Measuring Sense of Place: Methodological Aspects. *Tijdschrift voor Economische en Sociale Geografie* 96 (5): 467–476.

Starchenko, R. 2013. Dinamika Chislennosti i Rasseleniia Russkikh Kryma v XVIII-XIX Vekakh. *Vestnik KGU im. N.A. Nekrasova* 6: 38–41.

State Statistics Committee of Ukraine. 2004. All-Ukrainian Population Census 2001. [http://2001.ukrcensus.gov.ua/eng/results/].

Stokols, D. 1990. Instrumental and Spiritual Views of People-Environment Relations. *American Psychologist* 45 (5): 641–46.

Strong, M. 2018. Influence of Memory on Post-Resettlement Place Attachment. In *Explorations in Place Attachment*, edited by J. S. Smith, 19–32. New York: Routledge.

Teper, Y. 2016. Official Russian Identity Discourse in Light of the Annexation of Crimea: National or Imperial? *Post-Soviet Affairs* 32 (4): 378–396.

Thompson, B. 2016. Sense of Place Among Hunter-Gatherers. *Cross-Cultural Research* 50 (4): 283–324.

Uehling, G. 2004. *Beyond Memory: The Crimean Tatars' Deportation and Return.* New York: Palgrave Macmillan.

Virden, R. J., and G. J. Walker. 1999. Ethnic/Racial and Gender Variations among Meanings Given to, and Preferences for, the Natural Environment. *Leisure Sciences* 21 (3): 219–239.

Vodarskii, Y. E., O. I. Eliseeva, and V. M. Kabuzan. 2003. *Naselenie Kryma v kontse XVIII-kontse XX vekov: Chislennost', razmeshchenie, ethnicheskii sostav.* Moscow: Iri Ran.

Voitovich, L. 2009. *Formuvannia krims'kotatars'koho narodu.* Bila Tserkva, Ukraine: Vidavets' Oleksandr Pshonkivs'kyi.

Vozgrin, V. E. 1992. *Istoricheskie sud'by krymskikh tatar.* Moscow: Mysl'.

Williams, B. G. 1997. A Community Reimagined. The Role of "Homeland" in the Forging of National Identity: The Case of the Crimean Tatars." *Journal of Muslim Minority Affairs* 17 (2): 225–53.

_____ 2001. *The Crimean Tatars: The Diaspora Experience and the Forging of a Nation.* Boston, Mass.: Brill.

_____ 2015. *The Crimean Tatars: From Soviet Genocide to Putin's Conquest.* London: Hurst.

THE ROLE OF *NOSTALGIA* IN (RE)CREATING PLACE ATTACHMENTS FOR A DIASPORIC COMMUNITY

HOLLY R. BARCUS and AMANGUL SHUGATAI

The concept of nostalgia implies a level of romanticization of a place or time period, a selective "remembering" of a place held in the imagination of an individual or group of people. These imaginings produce a narrative that both extends place attachments to other community members, and potentially create shared myths of place. We use the Mongolian Kazakh population as a case study of how, over time, strong place attachments evolve into place-based narratives of home and identity for diasporic communities. We argue that strong place attachments are the foundation of place-based narratives, but these narratives are nostalgic as well, highlighting key ethnic and cultural practices that reinforce ethno-cultural unity and create idyllic images of place to help maintain a collective place valuing despite residence far away.
Keywords: Kazakh, mobility, Mongolia, nostalgia, rural idyll.

Imagining the "Nation"

> Endless open land stretches into apparent infinity across much of the Mongolian landscape ... Random *gers*,[1] roaming herds here and there, surrounded by nothing but nature's own sculptures, further contribute to the romantic notion that Mongolia is an unbounded land, where nomads wander free. (Myadar 2011, 335)

Orhon Myadar describes the images promoted by tourism agencies seeking visitors to Mongolia. Myadar continues, critically, to unpack these images, writing:

> The essentialised and romanticized construction of Mongolia as the 'timeless home of nomads' has endured – despite the fact that 'nomadic' hardly describes modern Mongolia or the majority of its people. Rather, the Mongolian landscape and Mongolian herders are used to construct and perpetuate the romantic, if medieval, portrayal of Mongolia in order to serve the need of outsiders for an imagined Other and the need of Mongolians for a cultural demarcation and social bond. (Myadar 2011, 336)

As Myadar so succinctly illustrates, the romanticized images of rural people and places helps establish narratives and imaginaries that can bring populations together —a nationalist trope. Myadar is speaking broadly of Mongolia and the imagery that is perpetuated about this rapidly developing and urbanizing place. These images stand in contrast to reports of life in Ulaanbaatar, where 45.7 percent of the

3.1 million population resides (NSOM 2016b, 27) and which Joseph Hincks gives the dubious distinction as "the most polluted capital in the world" (2018). Myadar's comments speak to the influence of imagery on national social imaginations and the portrayal of idealized, if highly unrealistic, images of place. But these images are grounded in a history that many Mongolians remember. Mongolia's rapid development moved the nation from a socialist society where 57 percent of the 2 million population resided in rural places in 1989 (NSOM 2003, 79) to one that is now 68 percent urbanized (NSOM 2016b, 50). But these images do not speak to all young Mongolians, who occupy a digital, globalized world, travel internationally, and seek the latest trends via their mobile phones, and who may see their history, but not their selves, represented in these nationalist rural images. As one young, affluent, Western educated Mongolian woman laments in flawless English, " ... they still don't have *real* coffee in the countryside ... " Although a seemingly offhand comment, it nonetheless underscores a tension between place imaginaries and place realities, and the evolution of place attachments at different spatial scales and across time periods and generations.

Much has been written about nationalism, ethnonationalism, and homelands as these concepts relate to the emergence of new nation-states in post-Soviet countries (Esenova 2002; Diener 2009; Tsuda 2009), as well as nationalism specifically in Mongolia (Batbayar 2002; Bulag 1998; Sneath 2010; Diener and Hagen 2013). The project of nation-building in Mongolia began during the Socialist period and was reoriented during the early transition years (from early to mid-1990s) when Mongolia shifted to a democratic form of governance and capitalist economy. Much of the imagery associated with Mongolia's nationalism reflect those images so aptly described by Myadar. These images depict landscapes and livelihoods that are common in Mongolia to the majority of the Khalkh ethnic group (84.5 percent in 2015), if not to the 22 other ethnic groups (NSOM 2016b, 36). The Kazakh population, however, has taken a different path. In the early 1990s, Kazakhs became the second nationality (*ündesten*) in Mongolia (Bulag 1998; Sneath 2010). This group, comprising 3.9 percent of the total population of Mongolia, is not only ethnically and culturally distinct from the dominant Khalkh ethnic group, but is largely clustered in western Mongolia, primarily Bayan-Ulgii *aimag* (province) and Hovd *aimag* (Figure 1). In Bayan-Ulgii, Kazakhs comprised nearly 88.6 percent of the total population in 2015, making the province an ethnic minority-majority area (NSOM 2016a, 28, 40). Mongolian Kazakhs are recognized by the state as a separate nationality and maintain an ethno-cultural territory in an isolated region of the country, where they maintain a strong presence, despite successive and sometimes extensive migration flows to Kazakhstan (Diener 2009; Werner and Barcus 2009) and more recently to Ulaanbaatar (Barcus and Shugatai 2018). Holly Barcus and Cynthia Werner (2015) argue that regardless of the ebb and flow of migrants from this region, Mongolian Kazakhs continue to maintain strong place attachments to western Mongolia. These place attachments take the form of religiosity, kinship ties, and language versatility—elements visible on the landscape in the form of mosques, and Kazakh language business signs (Barcus and Werner 2015). As migration trajectories shifted towards Ulaanbaatar in the later 2000s, many Kazakhs continued to remain in place, acting as cultural anchors of place attachment and ethnic identity for those who migrated away (Barcus and Shugatai 2018).

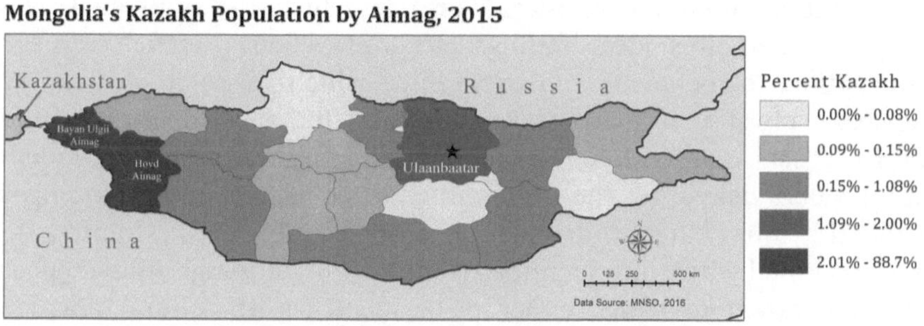

FIG. 1—Kazakh Population in Mongolia.

In this paper, we argue that the strong place attachments, developed in western Mongolia and perpetuated by Kazakh migrants to Ulaanbaatar, are evolving into a form of rural idyll, which is passed to younger generations through images of bucolic places, material reminders of place, and community festivals and celebrations. Kazakh migrants to Ulaanbaatar actively seek to keep these culturally specific rural images vibrant for younger generations, despite having no inclination towards return migration. We build our conceptual framework around three key literatures—nostalgia, place attachment and identities, and the rural idyll—in our attempt to better understand how connections to place evolve over time for populations who have moved away from the home place. Drawing from Leila Scannell and Robert Gifford's (2010) conceptual framework of a multidimensional place attachment, we build on the idea that place attachment is a process extending from the individual scale to broader community scales via social imagination and processes of memory and nostalgia. We use the Mongolian Kazakh population as a case study of how, over time, strong place attachments evolve into place-based narratives of home and identity for diasporic communities. In the next section we provide an overview of the literatures upon which we draw to develop our argument, followed in the third and fourth sections with our case study and methodology sections. Our analysis highlights our findings and the last section offers concluding thoughts.

Conceptual Framework: The Imagination of Place, Migration, and Nostalgic Rememberings

Place Attachment and Place Identities

Since Edward Relph's 1976 book, *Place and Placelessness*, highlighted the question of whether people were losing their connections to place, scholars have sought to define and understand the connection between people and places. Relph focused on concepts of insideness, outsideness, and authenticity of place. Taking a different track, behavioral psychologists, such as Proshansky and others (1983), consider place from the perspective of the self. For Harold Proshansky and others, "[Place Identity] is a sub-structure of the self-identity … consisting of … cognitions about the physical world in which the individual lives" … including " … memories, ideas, feelings, attitudes, values, preferences, meanings, and conceptions of behavior and experience which relate to the variety and complexity of physical settings" (1983, 59), basically place identity is about the self, a form of self-identity. John Harner, however, argues that "[p]lace identity is a cultural value shared by the community, a collective understanding about social identity intertwined with place meaning" (2001, 660). Culture, thus, is deeply intertwined with place identities (Relph 1976), although individual agency mediates the extent to which either place identities or cultural practices are internalized or practiced.

In 2010, in response to the growing number of definitions and dimensions within the place-attachment literature, Scannell and Gifford proposed a multifaceted definition for place attachment, which they termed the Person-Process-Place (PPP) Model (2010) (Table 1). "According to [the] person-process-place framework, place attachment is a bond between an individual or group and a place that can vary in terms of spatial level, degree of specificity, and social or physical features of the place, and is manifested through affective, cognitive, and behavioral psychological processes" (Scannell and Gifford 2010, 5). This framework is useful for laying out the nuances of place attachment, how at different scales place attachment can be both highly specific to an experience of one person while capturing a broader sense of connection for the group. It also helps us consider disentangling the existence of place attachments from the process of creating place attachments.

Evaluating the PPP Model and other definitions of place attachment, Bernando Hernández and others assert that much attention has been paid to defining and

Table 1—"The Tripartite Model of Place Attachment" (Adapted from Scannell and Gifford 2010, 2)

PLACE ATTACHMENT						
PERSON		PROCESS			PLACE	
INDIVIDUAL	CULTURAL/ GROUP	AFFECT	COGNITION	BEHAVIOR	SOCIAL	PHYSICAL
Experience Realizations Milestones	Religious Historical	Happiness Pride Love	Memory Knowledge Schemas Meaning	Proximity-maintaining Reconstruction of place	Social arena Social symbol	Natural Built

measuring the "person" aspect of the PPP Model, with more attention needed on the "process" and "place" dimensions (Hernández and others 2014, 134). While one aspect of process is how place attachments develop for either an individual or a group, another perspective is to consider how place attachments change through time. In other words, once one establishes that place attachments exist, do these same attachments remain static through time? Two recent dimensions of this research include the importance of "memory as an enabler of place attachment" (Lewicka 2014, 49) and the interplay between mobility and place attachment. As others have noted, there is much evidence that length of residence in a place affects place attachments, with place attachment often increasing with length of residence (Gustafson 2014; Lewicka 2014). However, as astutely pointed out by Maria Lewicka, although this association seems clear, it is of moderate strength with many additional factors clearly contributing to an overall sense of place attachment (2014, 50–51). Mobility and place attachment, however, are also not mutually exclusive; rather, many examples exist of high-mobility societies with strong place attachments (Gustafson 2014, 38).

Place attachments and identities, when viewed from afar, whether by geography or time, and especially when held in comparison to a challenging living environment in the present, may simultaneously take on positive images and lose negative aspects. The imagined home place takes on a mythical existence in personal and group narratives, complete with key social and familial practices, and bucolic images of peaceful, beautiful places. Such narratives can be specific to an individual, a group, or indeed a nation, as Myadar (2011) so aptly notes.

Linking Place Attachment, Place Elasticity, Nostalgia, and Migration

Migration away from a beloved home place, whether voluntary or forced, can create a desire to remain connected, however imperfect, through cultural practices, oral histories, or storytelling that evoke portrayals and imagery of a faraway place. Family histories and genealogies, for example, are a means by which individuals establish connections to the past, inclusive of people and places (Lewicka 2014, 53-54). The concept of place elasticity further describes how individuals who move away from a home place seek to maintain connections and identities associated with that place, including strong place bonds, permanence, and portability (Barcus and Brunn 2010). Place elasticity is premised on an individuals' connection to a place—that there is a reason to perpetuate connectedness between people and place. In a previous article, we argued that " ... immobile populations play a key role in maintaining and perpetuating cultural narratives for ethnic minority populations by anchoring cultural narratives to place" (Barcus and Shugatai 2018, 8). For example, places such as summer pastures, or local institutions, or burial sites might foster strong place bonds through direct experience with these places. Out-migrants may also maintain a continued dialog with a place (portability) by staying up-to-date on local news and events via social media or other telecommunications technologies, visiting, or attending family events. Lastly, permanence describes a " ... desire to perpetuate place bonds through time and generations ... ." (Barcus and Shugatai 2018, 8). In this article, we see place elasticity as establishing a connection between migrants and western Mongolia, and a broadly shared sense of identity specific to place. In other words, these are not just cultural or ethnic identity practices but rather they are cultural and ethnic identities that are associated with a specific territory or place that is at once experienced and reimagined through time. It is this aspect that we see connecting to concepts of nostalgia.

The concept of nostalgia implies a level of romanticization of a place or time period, a selective "remembering" or "creating" of a place held in the imagination of a person or group. Such rememberings or imaginings are held not only by individuals, but begin to produce a narrative that both extends place attachments to other community members and also potentially creates shared myths of place dependent as much upon selective forgetting as selective remembering. "Nostalgia is adaptive—it helps to put together broken parts, builds a bridge between past and present, increases self-esteem and life satisfaction, and reinforces social ties" (Lewicka 2014, 53). Nostalgia and place elasticity, then, find common ground linking people and places through memory and imagination.

While the concept of place elasticity helps us understand how people remain connected to a specific place after migration, the idea of nostalgia describes how memory and imaginaries interact to (re)create or preserve particular valued characteristics of a specific location, thus potentially transferring place attachments across generations.

CREATING A RURAL IDYLL?

Although discussed extensively in Global North studies of rurality (Bunce 1994), scholars have not sufficiently applied the concept of the rural idyll to Global South and Global Middle[2] countries. In the rush towards urbanization, modernization, and "development," sentiment for the rural and the old somehow seems to have escaped the scholarly gaze in these world regions. Or more likely, the quest for the modern, global urban lifestyles portrayed on television and the Internet can leave rural places seemingly distant and disadvantaged in comparison. But the view from within a population that has moved away may be quite different from outsider perspectives.

Writing of the origins of the rural idyll concept in his book *The Countryside Ideal*, Michael Bunce describes the link between "agrarian simplicity and rural nostalgia." "At heart it is an idea which romanticizes pre-industrial culture, casting the traditional rural lifestyle and communities of the past in nostalgic contrast to the dynamic individualistic culture of the present. Perhaps the most important element of this is the attachment of reverential status to farming as a way of life" (Bunce 1994, 29). Although, for Mongolia, we might substitute "herding" for "farming," Bunce captures the sentiment of nostalgia for rural places and the associated imagery of the rural idyll in North America. He further asserts that the theme of rural nostalgia is, historically, a relatively common " ... social response to the urbanization process ... " (Bunce 1994, 5). Key terms that embody the concepts of the rural idyll include "Nostalgic/part of national identity, traditional, problem free, closely knit/friendly, better environment, place for play, simpler/more natural," while the rural anti-idyll is described as "backward, unsophisticated, unfriendly, environmentally damaged, dull, boring, poorly provided with services, sleepy" (Yarwood 2005, 24). Each set of terms reflects a different perspective on rural. Richard Yarwood, however, is making a broader point, which is at least in part about how rural places are interpreted differently by different groups of people, and the images and associations with rural vary accordingly (2005, 21).

While nostalgia is a key aspect of the rural idyll (Yarwood 2005), linking nostalgia to national identity is also important for the development of a rural idyll. These linkages have been less common in Global South and Global Middle countries where the rush to "develop" often depicts rural places in terms similar to the rural anti-idyll. Mongolia, in this case departs from this portrayal of its rural spaces, at least on a national scale. Instead, bucolic rural images are the essence of nationalist images in Mongolia, despite scholarly critique (Batbayar 2002; Myadar 2011).

In summary, memory, nostalgia, and the narratives of place attachments and identity shared by migrants, with each other and across generations in the new destination, perpetuate a shared place attachment that reinforces social imaginations of a place. The following case study provides an example of this process. We argue that linking ideas of nostalgia, place attachment, and the rural idyll are increasingly important in Global Middle countries, where newly affluent urban populations may reflect on aspects of their rural origins that are less present in their urban, modern lifestyles. By using memory, knowledge, stories, narratives, artifacts, pictures, language, and urban festivals, former residents of western Mongolia create a "rural idyll" for younger generations. Western Mongolia becomes a social symbol of Kazakh culture and history—a rural idyll that, while specific to the Kazakh population, also embraces the images of landscape and rurality common to the broader nationalist images utilized by the Mongolian government.

Changing Mobilities and Place Attachments amongst the Mongolian Kazakhs

To develop our argument, we focus on recent urban in-migrants, whose ethnicity and cultural histories are closely tied to a particular remote, rural region of Mongolia, to explore the permutations of place attachment and identities as expressed through daily life. We further explore how these place attachments are communicated to younger generations, typically children, whose only personal experiences with the rural home of their parents or grandparents is through stories and short visits. In these ways and in everyday rememberings, place is invoked and created through memories and imaginations of others, thus creating in the social imagination of ethnic Kazakhs in Ulaanbaatar, a level of nostalgia for the rural places from which they or their kin migrated.

Who Are the Kazakhs in Mongolia?

Kazakhs comprise the second largest ethnic group in Mongolia (3.9 percent of the total population), after ethnic Khalkh Mongols, although the overall size of the Kazakh population is quite small (NSOM 2016b). From the perspective of cultural, linguistic, and religious practices, Kazakhs are distinct from the Khalkh Mongol majority, as most Kazakhs are Muslim, whereas Buddhism and Shamanism are more common among other ethnicities in Mongolia. In Bayan-Ulgii, while Mongolian is the language of interethnic communication and official business, ethnic Kazakhs primarily utilize Kazakh as their first language. In a relatively isolated and borderland province such as Bayan-Ulgii, this practice did not represent significant challenges to the Kazakh population and was one aspect of a strongly felt sense of ethnic identity in the region. The presence of mosques, especially in towns and in the provincial center Ulgii, further cemented a sense of place specific to this ethnic group (Barcus and Werner 2015). In essence, then, Bayan-Ulgii Aimag, despite being sparsely populated and on the borderlands of Mongolia, has a distinctively Kazakh human landscape; the mosques are both visible and audible, as the calls to prayer occur throughout the day. Kazakh is the primary language and celebrations, food ways, *gers* (portable, round dwellings), and handicrafts all reflect the influence of this ethnic group. Importantly, this is also a place in which ancestors are buried, and this is an important element of Kazakh place-based identities (Barcus and Werner 2015). The experience of having both a visible landscape reflective of the presence and material artifacts of the Kazakh population, as well as the ability to converse in Kazakh, creates a sense of place that is distinct culturally and linguistically from other places in Mongolia.

Movement away from western Mongolia can represent a significant disruption in an individual's sense of identity and place. It is not the intention of this research to ascribe identities and interpretations of individual identities to a group or individual, but rather to describe how living in this particular place with many coethnics can create a living environment and coethnic community that is not easily replicated elsewhere. It is also an important aspect of place elasticity and nostalgia—when an individual is far from a treasured place or home, elements of daily life may be ascribed more significance in remembering than might have been given originally.

The Case Study

The physical, cultural, and social landscapes of western Mongolia stand in stark contrast to those of Ulaanbaatar. As the primate city and capital of Mongolia, Ulaanbaatar is a sprawling, densely populated urban area with extensive ger districts, significant pollution issues, congestion, and social competition for employment and education. In our interviews, Kazakh respondents revealed levels of discrimination in access to services, as well as small acts of discrimination that make daily life much harder in Ulaanbaatar.

Internal migration from the countryside is a top challenge facing urban planners in Ulaanbaatar, including issues of overwhelming levels of air pollution, traffic congestion, and lack of infrastructure development (NCPH and UNICEF 2018). The city is divided into districts, including areas in which the residences are primarily apartments and those that are dominantly ger districts. Ger districts are areas of housing and commerce characterized by small, privately owned lots often utilizing the traditional felt ger. These districts lack water and sewer infrastructure and are largely unplanned and sprawling, although they include householders with a range of income and education levels. Nearly 60 percent of the population in Ulaanbaatar reside in the ger districts (NCPH and UNICEF 2018, 15).

Kazakhs comprise only a small portion of the overall migrant flow to Ulaanbaatar, are widely dispersed across the city, and compete with other migrants and residents for education, health care, and other basic social services and employment. While several districts are home to a larger number of Kazakhs, they are not significantly clustered into neighborhoods and therefore individuals must seek out coethnics through friend and family networks. A few large mosques exist in Ulaanbaatar (Figure 2), but they are distant from many Kazakh households so individuals must either travel greater distances or find alternative spaces to observe religious practices.

Methodology

We draw on 100 questionnaires and 36 in-depth life-history interviews with ethnic Kazakh migrants to Ulaanbaatar who previously lived in rural places, mostly in Bayan-Ulgii and Hovd aimags in western Mongolia. Questionnaires were collected during the summer months of 2015 and life-history interviews were conducted during the summer months of 2016. All data were collected in Ulaanbaatar. We chose participants using a nested quota sample incorporating age, sex, and migration experience, although we utilized a snowball sampling method to find potential participants. The majority of respondents migrated post-1989, with a small handful migrating pre-1989. Respondents have highly diverse levels of education, life experience, and human capital. Some migrated to Ulaanbaatar to attend university, others came with families to create better lives for their children, and others moved as professionals. The life-history interviews included a subset of the 2015 questionnaire sample as well as additional interviewees and are evenly divided between male and female respondents. About 33 percent are 18-34 years, 39 percent are 35–54 years, and 28 percent were 55 years and over. Questionnaires and life-history interviews covered a wide range of topics including demographic, occupational, and household information; migration histories and questions about migration; social and economic situations; access to services; and perspectives on cultural practices and life in Ulaanbaatar.

FIG. 2—Mosque in Ulaanbaatar. Summer 2016. Photo by Holly Barcus.

Our sample is intentionally divided to include respondents with different living experiences spanning the ger districts and apartments. We also added field notes and observations of the ways in which place attachments and nostalgia are expressed and intertwined in homes and daily lives. These include observations about how place is discussed, connections to place that may be displayed or evidenced in respondents' homes, and, occasionally, the foods served and discussed during our interviews. All life-history interviews and questionnaires were conducted by the authors, in Kazakh, Mongolian, or English, depending on the preference of the interviewee, although the majority were conducted in Kazakh. Interviews were transcribed and translated in English, and Atlas.ti was utilized to identify preliminary themes in the data. The authors are native speakers of Kazakh or English and one author is fluent in Mongolian. Interview location varied, although most interviews took place in interviewees homes, other locations included workplaces, parks, or cafés. In this article we primarily utilize the life-history interviews, although some questionnaire data are incorporated.

One advantage of this work is our long association with this population. While the research team has changed over the years, several of the original members remain part of the team. We, the authors, self-identify as women, and represent insider and outsider perspectives to this population. Engagement with this community includes multiple fieldwork periods in Bayan-Ulgii, including 2006, 2008, and 2009, as well as two fieldwork periods in 2015 and 2016 in Ulaanbaatar, and informal visits during 2014 and 2019. Migration trajectories have changed over these time periods, and it is not unusual to interview individuals in Ulaanbaatar during the later fieldwork periods (2015, 2016) that we interviewed in Bayan-Ulgii during the earlier fieldwork periods. We did not seek these individuals out specifically, but their insights over time illuminate variation in migration priorities that evolve at the individual level in response to policy changes and social and economic opportunities or challenges in both Mongolia and Kazakhstan over this extended period of time.

Finally, official statistics from the National Statistics Office of Mongolia (NSOM) allow for a tabulation of the total number of ethnic Kazakhs residing in Ulaanbaatar at different points in time. As we have noted in previous articles, the migration patterns of ethnic Kazakhs in Mongolia have shifted over time, favoring migration from Mongolia to Kazakhstan (and return) during the early and middle years of the economic transition (1990s to early 2000s) and then shifting towards Ulaanbaatar in more recent years. There are many reasons for the shifting mobility patterns, not least of all including practical citizenship considerations as well as more subjective imaginings of home and homelands (Werner and Barcus 2009; Barcus and Shugatai 2018). Interviews and ethnographies are important qualitative means of collecting information about place and migration and in particular, about the nuances of both the migration process, outcomes, and insights into home and place (McHugh and Mings 1996; McHugh 2000; Mendoza and Morén-Alegret 2012).

Analysis and Findings ~ Perpetuating Place Attachments

Our data and field notes reveal three dominant themes supporting the contention that place attachments are reimagined and (re)created by those who move away forming a nostalgic and often idyllic narrative of a rural home place for this small and dispersed population.

Bucolic Landscapes and Happy People

Telling and retelling of particular family and community narratives encourages a group-based nostalgia for western Mongolia. We situate this within the process-place dimensions of the PPP framework. The narratives highlight specific themes that underscore social and physical place characteristics that are difficult to replicate in Ulaanbaatar. They include narratives of close-knit family, friend networks, and reciprocal caretaking; ease of communication; a shared appreciation of religion, language, and foodways; and stories of nature, environment, and place. While these are descriptive of place, they are also essential to the process of reaffirming a particular place as important to this particular group of people. These portrayals, whether oral or symbolized in artifacts and practices, exemplify Yarwood's description of the rural idyll as, "Nostalgic/part of national identity, traditional, problem free, closely knit/friendly, better environment, place for play, simpler/more natural" (2005, 24).

For example, the first theme from our life-history interviews is the importance of the physical environment. We see this manifest not only in the interviews, but also in artifacts and images present in some of the interview spaces and the stories told by our respondents. Common elements of these stories include animals such as horses and sheep, gers, activities such as milking the animals, and natural features such as mountains and rivers. The environmental images are bucolic in nature, never indicating the harshness of the real environment, but rather focusing specifically on positive images, such as the beauty of the landscape.

Gulnar,[3] for example, is a 34-year-old mother of two with an economics degree from a Mongolian university. She and her husband travel widely due to their respective occupations, but own an apartment in Ulaanbaatar where we conducted the interview. Gulnar (LH_1_0016) is deeply concerned about the loss of Kazakh culture and the process by which they are educating their two young children. The family speaks only Kazakh at home, attends festivals when possible, and seeks to educate their children in Kazakh practices. When speaking about western Mongolia, she talks softly and thoughtfully, clearly drawing a picture for us of her experiences there. Her fondest memories include feeding and milking the animals and playing outside her family's ger. "I was too small for the horses but I learned how to milk the goats and the sheep the first time. It was really amazing … " (LH_1_0016).

Her memories of western Mongolia, like many other respondents, reflects the peak summer season experiences, when landscapes are beautiful and green, the weather is mild, and there is sufficient food for animals and humans (Figure 3). A middle-aged male, Bolatkhan, also from western Mongolia, draws a similar sense of longing for this place (LH_802_119): "My birthplace is very close with my heart. Sometimes I miss my birthplace. The beach by the river, herding livestock. I love them." He tells us that his most important reminders of western Mongolia are " ... the clean environment. The water is clear, there is also clean air. Here [in Ulaanbaatar] there is no fresh air. [In western Mongolia there are] cool and nice places." Both Gulnar and Bolatkhan tell these stories to us, but it is clear that these are also the images and stories frequently relayed to their children.

These memories and images of place describe the physical landscape and are deeply entwined with positive images of social and familial experiences. Alikhan, a 30-year-old male, for example, links the closeness of the *auls* (encampments of interlinked families in the countryside) to feelings of belonging and happiness. "The main difference is that in the small provinces we're close, you know everybody. So in that sense, I think that the atmosphere of life and the day-to-day life is that you can interact with everyone you know and say, 'Hi. How are you?' ... I kind of remember mostly playing together with the other kids. Activitywise, we went out in the morning and came back at night" (LH_3_0018). Again, these stories are frequently told to children about life in the countryside. For a widely dispersed urban population, these narratives portray Kazakh families as tightly linked and closely woven into the fabric of the broader community.

FIG. 3—Pastures in Western Mongolia. Summer 2009. Photo by Holly Barcus.

In each account, the respondent conjures images of bucolic rural landscapes, creating an idealized and nostalgic version of rural life, of life in the summer pastures. These images, however, are not stand-alone. Each also includes a social landscape that is equally harmonious. For example, Beibut, a 67-year-old male respondent living in the ger districts tells us of his childhood: "When I was young, I had three sisters and a little younger brother and I was the eldest boy in my family. When I was a little boy, I remember riding a horse. My father had many horses and he participated in horseracing. I enjoyed it very much. I was riding a horse still when I was 16 years old. In the summer my family moved [to the summer pastures] from Tsengal to Dayan. It was a nice place to camp. In the summer we usually milked the mares. All our relatives were living there. It was nice" (LH_19_0034). Our interview took place at his home in Ulaanbaatar, a compound we described in our notes as a bit worn, with two aging gers occupying the center of the area. Beibut, a retired scientist, worked with the Russians until 1990. He lived in Bayan-Ulgii until 1997 when he migrated to Ulaanbaatar, bringing his family to live with him later. On the day of our interview, he was dressed in white with an embroidered Kazakh *tebeteika* (an embroidered hat worn by Muslim men), surrounded by traditional handicrafts, including felted wool rugs (*syrmaqs*) with *tus kiiz's* (traditional embroidered wall hangings). Behind him were black-and-white pictures of family members, as well as medals for horse riding and eagle hunting. He is active in the Kazakh community and interested in preserving Kazakh culture. In this theme, the importance of both the environmental and social landscapes convey a sense of belonging and beauty ascribed to western Mongolia. These images stand in contrast to experiences in Ulaanbaatar, where Kazakhs are rarely able to converse in Kazakh, except at home, and the environment is more urban and industrial.

Material Reminders of Place: Textiles and Artifacts

A second theme linking place attachment to nostalgia and the creation of a rural idyll is the presence of material artifacts in many homes. Traditional textiles are often subtly displayed, integrated amongst everyday items, and used for daily activities. For example, in many kitchens where we conducted our interviews over tea and *baursak* (a type of fried bread), the small kitchen tables were covered with pieces of Kazakh embroidery overlain with clear plastic for protection. Other textiles were often visible in homes; textiles that are clearly not made for general consumption, but rather either produced by or for the household specifically.

Syrmaqs are felted wool rugs (Figure 4), handmade by women and often given to a young couple as a marriage gift from the bride's family (Portisch 2009). Although we never witnessed syrmaq production in Ulaanbaatar, it was clear that households used them for various purposes. In one home, Ainur, a middle-aged woman, was caring for a group of young children. As we conducted the interview, the children were settled in an adjoining room watching cartoons on television and having a snack while sprawled on several layers of overlapping syrmaqs. Portisch writes about how textiles, and syrmaqs in particular, are often incorporated into the daily living and life lessons of children.

> For a child learning about syrmaq, this process may begin in a number of different situations. The child may begin to learn to recognize and remember this artifact as a woolly surface that is rough and itchy to sit on against the bare skin. It may be warmer sitting on a syrmaq than elsewhere on the floor, and the child may have a visual impression of the colours and patterns as well as an olfactory impression of the syrmaq, which smells of the unwashed lamb's wool it is made of. Later, the child might know this type of artefact as an object that has been designed and created by her mother or grandmother, and is used in certain situations. (Portisch 2009, 476–477)

Syrmaqs, tus kiiz's (Figure 5), and other embroidered textiles did not dominate living spaces, but were subtly integrated with more contemporary furnishings. They nonetheless point to a connection between place and community. Discussing craft production in Bayan-Ulgii, Guldana Salimjan writes that "[j]ust as falconry became a vital connection to the Kazakh historical past, natural environment, and traditional culture across Eurasia, handicrafts have also allowed Kazakhs to maintain their identity and traditional knowledge as an ethnic minority in Mongolia, and have become an attribute of 'authentic' Kazakhness" (2018, 2). Unlike falconry, which takes on a somewhat iconic symbolism, crafts, such as syrmaqs and tus kiizs, are everyday reminders of ethnic identity and place connection. Handicrafts are dominantly produced in households in rural places and they are unique to the individual women who produce each item. Typically this is a family member. Tus kiiz, for

FIG. 4—Kazakh Syrmaq. Summer 2009. Photo by Holly Barcus.

example, may be passed from mother to eldest daughter at marriage, thus creating another familial link to place and identity, as, in these cases, the textiles were produced by women in western Mongolia and passed through the generations—a further reminder of the importance of place passed along between generations.

Salimjan writes that "[t]us kiiz is a medium for such traditional knowledge transmission as well as identity formation … Material culture speaks to embodied memories around interpersonal relations and people's connections to land and the materials around them" (2018, 14–15). Finding tus kiiz in our respondents' homes, and additionally, similar embroidered pieces that are reimagined into everyday items, such as tablecloths, reflects the connections that Kazakhs residing in Ulaanbaatar still wish to maintain to western Mongolia. Such material reminders of place underscore the reflective, nostalgic aspects of place attachment for migrants living far from western

FIG. 5—Kazakh Tus Kiiz. Summer 2009. Photo by Holly Barcus.

Mongolia, acting as reminders of place and home and continuing to connect migrants to a distant but important place. They become the everyday items that younger generations now also see as part of their connection to this faraway place.

The Practice of Place: Festivals, Community, and Language

The third theme emerging from the interviews is the role of food, festivals, and language in perpetuating place attachments and cultural identities. Festivals and festival attendance are one manifestation of culture on the landscape. Two large and widely celebrated events for Kazakhs in Mongolia include the Eagle Hunting Festival(s) and *Nauryz* (the Kazakh New Year celebration). Family events such as the birth of a child, weddings, and funerals are also important festive gatherings of kin. As one respondent notes, "It is very difficult to keep our religion and culture and tradition. We are a small ethnic group in Mongolia. So we often organize Kazakh religious and cultural events" (LH_36_0051). From our questionnaires, we know that 82 percent of respondents attended at least one Kazakh festival in Ulaanbaatar in the year preceding the interview and only 2 percent had no knowledge of Kazakh festivals (Barcus and Shugatai 2018).

Celebrations, whether public or family-based, create an atmosphere of shared culture and reinforce place attachments and cultural identities across generations. For example, Nauryz is publically celebrated but also has an after-party among smaller kin and friend groupings of Kazakhs. Gulnar, a 34-year-old female respondent tells us, "Yes recently I attended Nauryz . . . All the Kazakhs in Ulaanbaatar gathered there and they also gathered together [after the official celebration] and celebrated Nauryz together in a big hotel and had a ceremony" (LH_1_0016). The Eagle Hunting Festival, of which there are several, including one in Ulaanbaatar and one in Bayan-Ulgii, also provides a level of public pride in cultural identity. *The Eagle Huntress* movie released by Sony Pictures in 2016, for example, displayed to the world the unique cultural practice of falconry with golden eagles that is tied specifically to the Kazakh population and now, through this mass media form, to western Mongolia specifically. Another way in which Kazakhs evoke images of western Mongolia at these public and private gatherings or celebrations is through music and musical instruments. Jennifer Post argues that musical expression, particularly with " . . . the *dombra*, a long-necked lute widely considered a symbol of Kazakh identity . . . ", is utilized by Kazakhs as a component of place-making in a community context (2007, 47). There are also NGOs that provide Kazakh language instruction, dance and dombra lessons for children, and a "Kazakh Students Bolashak Association" in several universities in Ulaanbaatar. Children may attend after-school or weekend "cultural classes" to learn Kazakh dance, music, and language, performing later at recitals or Kazakh festivals. In these ways, place attachment and cultural identities are perpetuated across generations through performance and public group affirmations.

Lastly, while most questionnaire respondents (87 percent) reported speaking Kazakh at home at least sometimes, gatherings provide a community-based forum for collectively speaking Kazakh, a sense of unity that is not possible when engaging with non-Kazakhs in daily life. Language and its perpetuation come up in every life-history interview—the importance of maintaining the Kazakh language, of speaking it at home and teaching children, and of the desire for Kazakh language and cultural schools in Ulaanbaatar are common threads. Encouraging children to learn Kazakh is significant to many families, even those that noted how important fluency in Mongolian or English, and sometimes Russian, was for social mobility in Ulaanbaatar. Some Kazakh migrants see a lack of fluency in Mongolian, or speaking with an accent, as an impediment to finding employment or doing well in Ulaanbaatar, where the remainder of the population speaks Mongolian. Interviewees perceive speaking English as essential for international business, and to some extent Russian is still valued, albeit distantly to Mongolian and English. Memories of western Mongolia center on the ease of language and communication and not needing to worry about speaking Mongolian. By extension, there is also an underlying fear that the children will not learn to speak Kazakh and so the language will be lost.

By using memory, knowledge, stories, narratives, artifacts, pictures, language, and urban festivals, former residents of western Mongolia create a "rural idyll" for younger generations. These are integrated into daily living as well as through events such as festivals. Western Mongolia becomes a social symbol of Kazakh culture and history—a rural idyll for this small population. Children are active participants in these activities and practices, absorbing, as children do, their parents and grandparents attitudes. Within the PPP model, this connects directly to ideas of cognition (memory, knowledge, and meaning) and affect (happiness, pride, and love) [Table 1].

For many Kazakh children, these three threads come together during summer holidays when families and sometimes only the children spend time in western Mongolia with relatives. "In summer vacation [children] should go to their grandparents [in the countryside] … .it is important to keep language and religion and respect our language and religion" (Batima, age 22, female, LH_25_0040). In the countryside, children are further exposed to traditional lifeways as these visits most often involve extended stays in summer pastures, living in traditional gers, and engaged in daily practices of milking, food preparation, horse riding, and living amongst an extended family. Children are fully immersed in the Kazakh language. Families generally return to the same pastures each year and this becomes another aspect of inscribing place and identity through everyday practices for children who ultimately will return to the city.

It is important that these place attachments and their perpetuation, however nostalgic, also reflect the types of nationalist images that Myadar (2011) depicts in the opening paragraphs of this paper. These images and narratives have a specific Kazakh flare, exemplifying Kazakh culture and practices. Post argues that "[n] ational markers (including language, food, clothing, social patterns, religion, musical instruments and musical expression) reinforced nationally and internationally have few connections to the everyday lives of the Kazakh people" (2007, 47). Belonging to the Mongolian nation is an important aspect of Kazakh existence in Mongolia, but so too is the specificity of this particular region of western Mongolia. The two sets of images align generally, but the place attachments expressed by Mongolian Kazakhs are much more specific to their cultural and ethnic identity, inclusive of language and religious practice, and to the western Mongolia communities in which the social networks and social ties also complement the rural imagination.

Concluding Thoughts

This work suggests that the process by which individuals and groups selectively develop narratives of history and place serves to underscore a collective valuing of place through signifiers of place attachment, and illustrates how the process of creating place attachments manifests at both individual and collective scales. We contribute to the discussion of how rural place attachments evolve over time and that place attachments can be as much symbolic as experienced. There is also a generational dimension to these rememberings, where those who grew up in western Mongolia offer nostalgic, bucolic images of life on the steppe to younger Kazakhs, thereby perpetuating positive imagery and imaginings of western Mongolia. The younger respondents offer equally thoughtful and positive images of western Mongolia, but are clear that life there would not be desirable. Thus, life in western Mongolia, for the Kazakh ethnic minority population living in Ulaanbaatar, is evolving into something of a mythical place—a place where the air is clear, the scenery is beautiful, where all Kazakhs take care of one another, and where Kazakhs collectively speak their own language and practice their religion. In the absence of discussion about the challenges of poverty and environmental hazards (*dzud*), the narrative has evolved into one of nostalgia—a set of images and ideals that are now passed along to a much younger generation, through small everyday practices and experiences as well as through attending and participating in festivals and summer holidays spent in western Mongolia. Bayan-Ulgii, with its majority Kazakh population, is likely to live in the memories and imaginations of Mongolia's Kazakh population as this mythical place, despite many of its more-harsh economic and environmental realities.

Our work underscores how place attachments are carried with migrants to new destinations (place elasticity) and then passed along to younger generations through stories, material artifacts, cultural practices, and images. This contributes to our understanding of the process and place dimensions of place-attachment literatures as proposed by Scannell and Gifford (2010). We also pose the question of whether the concept of the rural idyll can be applied to Global South and Global Middle countries. So far, the concept has largely been reserved for the Global North. But as populations continue to move to urban places, can we expect that the rural imagination will expand? That instead of being viewed negatively, as the place left behind, there will be elements of nostalgia and positive imagery shaping the narrative of particular populations, especially ethnic minorities who find themselves widely dispersed after migrating to urban places.

We move the literature forward by seeking to integrate the notion of nostalgia into the conversation about rural place-making and place identities or attachments. Nostalgia exemplifies the positive aspects of place and pushes us to consider how rural imagery and imaginations influence the place-making and place-attachment process, which begs the question of whether the rural idyll concept can be extended beyond the Global North. Rural Mongolia, with its wide-open spaces, green rolling grasslands, horses, gers, and blue skies inhabits the imagination of tourists and is a key component of nationalist images promoted by the Mongolian government. It is, however, unusual to suggest that a rural idyll has emerged in a Global Middle country. We argue that while the urban elite of Mongolia still view the rural as undeveloped, the narrative of the idyll is indeed emerging for smaller, ethnic minority populations for whom the urban environment is often alienating and hostile. Nostalgia for the simplicity of a remote, rural territory takes on greater meaning. It is romanticized and narrated to younger generations, much like the rural idyll, but not with the expectation of return migration.

Funding

This work was supported by the Andrew W. Mellon Foundation [New Directions Fellowship].

Notes

[1] "A portable round dwelling, typically made of felt and wood" (Myadar 2011, 357).

[2] Mongolia is considered a Global South country as well as a "Lower Middle Income Country" by The World Bank (World Bank, 2020, https://data.worldbank.org/?locations=MN-XN).

[3] We use pseudonyms to protect interviewee identities.

ORCID

Holly R. Barcus http://ORCID.org/0000-0002-3841-5725

References

Barcus, H. R., and S. D. Brunn. 2010. Place Elasticity: Exploring a New Conceptualization of Mobility and Place Attachment in Rural America. *Geografiska Annaler: Series B, Human Geography* 92 (4): 281–295.

Barcus, H. R., and C. Werner. 2015. Immobility and the Re-imaginings of Ethnic Identity among Mongolian Kazakhs in the 21st Century. *Geoforum* 59: 119–128.

Barcus, H. R., and A. Shugatai. 2018. Immobile Populations as Anchors of Rural Ethnic Identity: Contemporary Kazakh Narratives of Place and Migration in Mongolia. *Population, Space and Place* 24 (4): DOI: 10.1002/psp.2148

Batbayar, T. 2002. Geopolitics and Mongolia's Search for Post-Soviet Identity. *Eurasian Geography and Economics* 43 (4): 323–335.

Bulag, U. E. 1998. *Nationalism and Hybridity in Mongolia*. New York: Oxford University Press.

Bunce, M. 1994. *The Countryside Ideal: Anglo-American Images of Landscape*. London: Routledge.

Diener, A. 2009. *One Homeland or Two? The Nationalization and Transnationalization of Mongolia's Kazakhs*. Stanford, Calif and Washington D.C.: Stanford University Press and Woodrow Wilson Press.

Diener, A., and J. Hagen. 2013. From Socialist to Post-Socialist Cities: Narrating the Nation through Urban Space. *Nationalities Papers: The Journal of Nationalism and Ethnicity* 41 (4): 487–514.

Esenova, S. 2002. Soviet Nationality, Identity and Ethnicity in Central Asia: Historic Narratives and Kazakh Ethnic Identity. *Journal of Muslim Minority Affairs* 22 (1): 11–38.

Gustafson, P. 2014. Place Attachment in an Age of Mobility. In *Place Attachment: Advances in Theory, Methods and Applications*, edited by L. C. Manzo and P. Devine-Wright, 37–48. New York: Routledge.

Harner, J. 2001. Place Identity and Copper Mining in Sonora, Mexico. *Annals of the Association of American Geographers* 91(4):660–680.

Hernández, B., M. C. Hidalgo, and C. Ruiz. 2014. Theoretical and Methodological Aspects of Research on Place Attachment. In *Place Attachment: Advances in Theory, Methods and Applications*, edited by L. C. Manzo and P. Devine-Wright, 125–137. New York: Routledge.

Hincks, J. 2018. Life in the Most Polluted Capital in the World. *Time*, 23 March 2018. https://time.com/longform/ulan-bator-mongolia-most-polluted-capital/Last accessed 1/8/2020.

Lewicka, M. 2014. In Search of Roots: Memory as Enabler of Place Attachment. In *Place Attachment: Advances in Theory, Methods and Applications*. Edited by L. C. Manzo and P. Devine-Wright, 49–60. New York: Routledge.

McHugh, K. 2000. Inside, Outside, Upside Down, Backward, Forward, Round and Round: A Case for Ethnographic Studies in Migration. *Progress in Human Geography* 24 (1): 71–89.

McHugh, K. E., and R. C. Mings. 1996. The Circle of Migration: Attachment to Place in Aging. *Annals of the Association of American Geographers* 86 (3): 530–550.

Mendoza, C., and R. Morén-Alegret. 2012. Exploring Methods and Techniques for the Analysis of Senses of Place and Migration. *Progress in Human Geography* 37 (6): 762–785.

Myadar, O. 2011. Imaginary Nomads: Deconstructing the Representation of Mongolia as a Land of Nomads. *Inner Asia* 13 (2): 335–362.

NCPH [National Center for Public Health] and UNICEF. 2018. Mongolia's Air Pollution Crisis: A Call to Action to Protect Children's Health. Mongolia: Ulaanbaatar.

NSOM [National Statistical Office of Mongolia]. 2003. Mongolian Population in XX Century. National Statistical Office of Mongolia. Mongolia: Ulaanbaatar.

———. 2011. N Mongolia National Census 2010 Provision Results. Archived September 15, 2011 (in Mongolian).

———. 2016a. *2015 Intermediate population and housing census*. Mongolia: Ulaanbaatar. (in Mongolian).

———. 2016b. *2015 Population and Housing By-Census of Mongolia: National Report*. Mongolia: Ulaanbaatar.

Portisch, A. O. 2009. Techniques as a Window onto Learning: Kazakh Women's Domestic Textile Production in Western Mongolia. *Journal of Material Culture* 14 (4): 471–492.

Post, J. C. 2007. 'I take my Dombra and Sing to Remember My Homeland': Identity, Landscape and Music in Kazakh Communities of Western Mongolia. *Ethnomusicology Forum* 16 (1): 45–69.

Proshansky, H. M., A. K. Fabian, and R. Kaminoff. 1983. Place-Identity: Physical World Socialization of the Self. *Journal of Environmental Psychology* 3: 57–83.

Relph, E. C. 1976. *Place and Placelessness*. London: Pion.

Salimjan, G. 2018. The Afterlives of Yurt Wall-hangings: Tus Kiiz. The CESS Blog: A Blog of the Central Eurasian Studies Society. [thecessblog.com/2018/07/the-afterlives-of-yurt-wall-hangings-tus-kiiz-by-guldana-salimjan-university-of-british-columbia/].

Scannell, L., and R. Gifford. 2010. Defining Place Attachment: A Tripartite Organizing Framework. *Journal of Environmental Psychology* 30: 1–10.

Sneath, D. 2010. Political Mobilization and the Construction of Collective Identity in Mongolia. *Central Asian Survey* 29 (3): 251–267.

Tsuda, T. 2009. Conclusion: Diasporic Homecomings and Ambivalent Encounters with the Ethnic Homeland. In *Diasporic Homecomings: Ethnic Return Migration in Comparative Perspective*, edited by T. Tsuda, 325–350. Stanford, Calif.: Stanford University Press.

Werner, C.A. and H.R. Barcus. 2009. Mobility and immobility in a transnational context: changing views of migration among the Kazakh diaspora in Mongolia. *Migration Letters* 6 (1): 49–62.

Yarwood, R. 2005. Beyond the Rural Idyll: Images, Countryside Change and Geography. *Geography* 90 (1): 19–31.

MULTI-SCALAR TERRITORIALIZATION IN KAZAKHSTAN'S NORTHERN BORDERLAND

ALEXANDER C. DIENER

ABSTRACT. Terms used in the study of people/place bonding too often circumscribe thinking to either the most readily conceived and tangibly experienced scale of place – home/neighborhood – or to the most politically prevalent scale of place – the state. In doing so, researchers ignore the influence of places that help constitute identity between and above the aforementioned scales. A more elastic conceptualization of place attachment allows for mutability of place-identities and their capacity to be both multi-sited and varying in scale. With a long history of ethnic Russian habitation from the Tsarist through Soviet eras, the northern oblasts of Kazakhstan emerged as a borderland of potential irredentism akin to eastern Ukraine. Calls for secession and annexation have, however, been minimal thus rendering the region's amalgam of place identities a ready case for considering relational geographies of borderland place-attachment. *Keywords: Borderland, identity, homeland, kazakhstan, place attachment, Russian, territorialization.*

The most intense forms of place attachment commonly center on the abstract concepts of home and homeland (Antonsich 2010). But while considerable research has been conducted on domicile/neighborhood (Lewicka 2011), and ethnicity/nationalism literatures emphasize the significance of territory, far fewer studies address multi-scalar and multi-sited geographic belonging (Herb and Kaplan 2018). The tendency to use the term 'place attachment' for individuals/ small groups and 'territorialization' for larger groups risks circumscribing the topic to distinct scales thus ignoring other places pertinent to identity formation and maintenance.

How distinct are the two concepts? As defined by Alisdair Rogers et al. (2013) territorialization refers to the "organization of human activities by fixing them in spatial territory." Such territories are commonly the context in which place-identities are manufactured to reflect geographic belonging. Place attachment is defined by James Ponzetti (2003, 1) as involving "the elaborate interplay of emotion, cognition, and behavior in reference to place." These allied concepts are coterminous to the extent that they revolve around place meaning and people's commitment thereto. This article seeks to explore the intersections of these concepts and expand the capture of place attachment research, while also drawing attention to the concept's elasticity and role in forming new identities.

Rather than being defined by nested de jure spatial structures commonly associated with and sanctioned by states (i.e., municipality, county, state, province, department, *oblast*, *raion*, etc.), both individual and group place attachments are mutable and manifest a spectrum of intensity relating to multiple locations and varied scales of place. These include places so meaningful as to be considered sacred and irreplaceable (Verdery 1999), places of moderate personal investment (Malpas 2018), places of negative attachment (Kirkness and Tijé-Dra 2017), and even bonds to remote geographic imaginaries of personal, historical, or cultural significance (Hediger 2019). Perceptions of belonging and alienation develop in conjunction with this spectrum and form the building blocks of place-identities that are more mutable than commonly believed.

In the wake of USSR's collapse in the early 1990s, after Russia's invasion of South Ossetia in 2008, and following Russia's annexation of Crimea in 2014, Kazakhstan's northern borderland became a region of prospective secession and irredentism (Diener 2015). This relates to its large population of ethnic Russians dating to the Tsarist settlement of the steppe by incrementally emancipated Serfs during the 18th and 19th centuries, as well as, additive-migrations of Slavic and other ethnic peoples during the early to mid-Soviet periods.

With a population comprised of only 39.7% ethnic Kazakhs in 1991, Kazakhstan's leadership confronted a complex socio-political setting in which to build a state (see Table 1). Following the pattern of all former Soviet Socialist Republics, ruling elites chose an ethnic state name (Kazakhstan – i.e., "land of the Kazakhs"), thus signaling primacy of the titular community. Both ethno-national and civic national policies and

TABLE 1—KAZAKH AND RUSSIAN POPULATIONS IN NORTHERN OBLASTS OF KAZAKH SSR 1989*, 1999**, 2009***

	1989		1999		2009	
OBLASTS	KAZAKHS	RUSSIANS	KAZAKHS	RUSSIANS	KAZAKHS	RUSSIANS
Akmola	–	–	313488	329454	349086	264011
Aktobe	406823	173281	482285	114416	601463	292572
Atyrau	–	–	391672	38013	465033	33617
East-Kazakhstan	253706	613846	743098	694705	781732	561183
Guryev	338998	63763	–	–	–	–
Kostanay	183590	492237	314801	430242	328936	380599
North-Kazakhstan (w/Kokshetau 1989)	111631	372263	214697	361461	198641	200849
Pavlodar	268512	427658	311862	337924	353711	287970
Semipalatinsk	432763	300533	–	–	–	–
Tselinograd	193703	400941	–	–	–	–
Turgay	126975	87381	–	–	–	–
Uralsk	351123	216514	–	–	–	–
West Kazakhstan	–	–	399030	174018	432517	135813

*Source: *Vsesoiuznaya Perepis Naseleni 1989 Osnovie Itogi*, Moskva: Rossiiskaya Akedemiya Nauk 1990.

**Source: *Perepis Naselennya Kazakhstana 1999* Almaty: Agenstva Respublika Kazakhstana po Statitika 1999.

*** Source: *Kratkie Itoki, Perepis Naseleni 2009*, Almaty: Agenstva Respublika Kazakhstana po Statitika 2009.

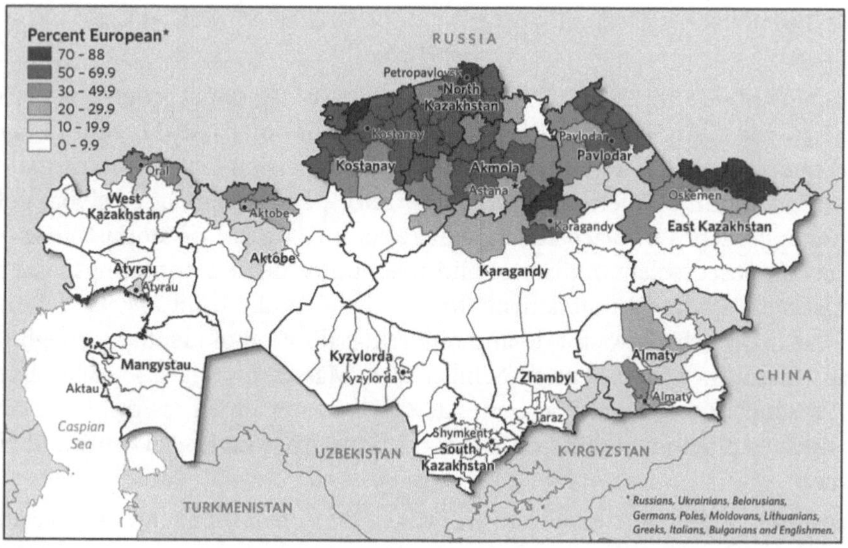

FIG. 1—Kazakhstan Percent European 2009. Source: *Kratkie Itoki, Perepis Naseleni 2009*, Almaty: Agenstva Respublika Kazakhstana po Statitika 2009. University of Kansas Cartographic Laboratory.

rhetoric have been advanced by the long-ruling Nazarbayev (1991-2019) and now Tokayev regimes (Fergana News 2020). The result is a rather ambiguous national identity that despite a 64 percent ethnic Kazakh majority by 2009 has generally checked extremism (Rees and Webb-Williams 2017). Whilst Kazakhstan's southern borderland possesses considerable numbers of Uzbeks, Kyrgyz, and Tajiks and the city of Almaty has a large Russian population, the northern oblasts constitute the Kazakhstani leadership's foremost concern pertaining to territorial integrity and efforts at nationalizing non-titular citizens (Diener 2015; Laruelle 2018; Cheskin and Kachuyevski 2019; see Figure 1).

Russians residing in this region are territorialized as a borderland population within a relational geography best captured by the Russian term *rodina* (Jašina-Schäfer 2019). This article seeks to explore their negotiation of multi-scalar place attachments that include membership in the Russian World, a civic national "Kazakhstani-ness," and personal geographic networks of daily spatial-experience. Subsequent sections of this article include a brief socio-political sketch of contemporary northern Kazakhstan, a methodology of data collection, analysis of data, and a conclusion pointing to future avenues of research on the topic. I begin by outlining how relational geographies can reflect multi-scalar aspects of place and identity.

Relational Geographies of Place Attachment, Sense of Place, and Place Identity

Approaching place attachment from the perspective of relational geographies emphasizes the co-constitution and mutual entanglement of otherwise different (and often distant) phenomena. Akin to John Agnew's (2009) assertion that the "territorial trap" limits consideration of alternative processes, actors, and dynamics of global politics, relational approaches to place attachment reject a Cartesian or atomist view of the world as composed solely of discrete and fixed units, de jure borderings, and time-embeddedness. Eliding reductionism to a strictly localized/intra-state definitions, a relational approach allows for de-and re-territorialization of identity through trans-national/external networks of place (Schiller 2005; Massey 2006). It also seeks causation beyond a single locus, regarding elements of place attachment as dispersed among interconnected phenomena (nature, social relations, meaning) (see Diener and Hagen this issue).

Place attachment/territorialization evokes two prominent forms of relational thinking in human geography. One accents overarching processes and relations that unify people into more or less stable place identities, while remaining affected by networked, dynamic, and horizontal connections across space (Massey 1993). The other attends to the experiences and actions of individuals, with a significant emphasis on difference and the possibilities of 'becoming' (Doel 2007, 810) This more elastic conceptualization of place attachment allows for the mutability of place-identities, deconstructs the dualism of home and homeland, and emphasizes the politics of relationality in which multiple sets of relations exist in both cooperation and competition (Murdoch 2006, 20).

"Borderlands" are productively approached through a relational lens, in that political actors generally function through territorial structures (Jones 2009; Paasi 1995). But rather than immutable entities, territories are made and remade through political, economic and cultural practices. Borderland-place-attachments are, therefore, sociospatial relations and practical outcomes of "a mutually transformative evolution of inherited spatial structures and emergent spatial strategies within an actively differentiated continually evolving grid of institutions, territories, and regulatory activities" (Jones 2009, 497-498).

Borderland studies considers place attachment as both de jure practice and de facto relationality that includes "sense of place," as a grounded lived-experience, and "place identities," as imposed, created, and manufactured perceptions of geographic belonging. Borderlands integrate both as multi-sited relational networks that complicate identities as territorialized through delimited geographies (i.e., state and its borders).

In contrast to the reductionist tendencies of the territorial trap, the very idea of 'borderlands' emphasizes place attachment as contingent, wherein positive or negative events can shift ideals of belonging away from less secure, less comfortable socio-spatial contexts (Anthias 2013, 7). Relational geographies are thus valuable for understanding how place attachment varies in intensity and can result in parochialism, inertia, movement, or entropy. By incorporating perceptions of relative social and/or economic deprivation/advantage as relational contingencies, place attachment fruitfully conjoins well-developed literatures on migration, diaspora, secession/irredentism, and transnationalism.

While Rogers Brubaker's (2011) warning against approaching identity through the binary of "ethnic" versus "civic" nationalism is well-heeded from the perspective of understanding personal identity, the tendency remains for post-Soviet states to deploy *jus sanguinis* perspectives on citizenship (Shevel 2019). Kazakhstan, by contrast, is among the region's foremost states enacting policies attuned to *jus soli* ideals. As shown below, this has allowed for re-territorialization of non-titular peoples and problematizes the definition of the state's ethnic minorities as 'diasporas.'

Understanding the capacities for non-titular groups to reterritorialize their respective identities is essential for the future of the region (Megoran 2017; Trimbach and O'Lear. 2015; Kaiser and Nikiforova 2006; Chinn and Kaiser 1996). Toward this end, Alina Jašina-Schäfer (2019, 2-3) argues for the broad semantic and cultural range of the Russian term *rodina*. Her definition of the term avoids the aforementioned conceptual and scalar circumscription of place attachment by rejecting 'native land' as precluding the possibility of an adopted *rodina* and demurring 'native country' or 'homeland' as too political and risking overlap with Russian words such as *otechestvo* (fatherland). Instead, she contends that *rodina* derives from the root '*rod*' (origin, species) and thus connotes an elastic notion of familiarity, partiality, and sense of ownership.

While possibly idiomatic and untranslatable into other languages, the term succeeds in capturing both the nebulous notion of geographic belonging and mutable aspects of both sense of place and place identity too often missing in studies of territorialization. *Rodina* is particularly useful for this study because (1) it is the term most commonly used in discussion of place attachment amongst Russophone peoples; (2) it encapsulates the multi-scalar spatiality extending from home to homeland; and (3) it attends to a relational approach to place attachment wherein localized place identities incorporate imagined communities of civilizational scale.

Brief Political Sketch of Northern Kazakhstan

The presence of non-titular populations across post-Soviet states has, in certain cases, manifested in ethnic conflict, irredentist tension, and annexation. The most prominent example is Ukraine, in which the popular geographic imaginary of a divided Ukraine (Riabchuk 2015; Charron 2016) has emerged in relation to Putin's claims for a *Novorossiya*, i.e., New Russia (Laruelle 2016b). Though some irredentist claims have been leveled at northern Kazakhstan, a scenario similar to Ukraine is unlikely for several reasons (Diener 2015; Laruelle 2018). First, there is no precise date or pivotal event with which an 'incorrectly' delineated border is associated.[1] Second, the region lacks a major city holding broad cultural or historical meaning within Russian culture.[2]

Finally, there is no 'natural' border between Siberia and Kazakhstan.[3] While some Russian irredentists have called for the whole of Central Asia to be reincorporated into Russian sovereignty (Zhirinovsky 2014), most limit their claims to northern Kazakhstan, but even they struggle to formulate a consistent historical or geographical claim (Laruelle 2018, 69).

Calls for separatism and/or irredentism from within northern Kazakhstan have also been rare. In the early 1990s, the organizations *Russkia obschina* and *Lad* were the primary fora of Russian public sentiment and various Cossack organizations occasionally spoke openly about sectioning off portions of the territories abutting Russia (Peyrouse 2007). But inconsistency reigned amongst grassroots activists. *Russkia obschina* and *Lad* proved incapable of coordinating their agendas, with the former joining the governmental advisory council, the Assembly of the Peoples of Kazakhstan, and the latter declining in influence along with other opposition parties during the 2000s (Ingram 2001; Laruelle 2015). Of Kazakhstan's three main Cossack groups, since the early 1990s, only the Uralsk group avoids participation in Kazakhstani state-sponsored activities, preferring interactions with Cossack *stanitsas* across the border in Russia. The other two Cossack groups (the Siberia and Semirechie) have demonstrated willingness to cooperate with the Nur-Sultan (née Astana) regime (Laruelle 2018, 73).

In contrast to Crimea and Eastern Ukraine, Kazakhstan's Russians tend to regard themselves as 'non-autochthonous.'[4] Far from discounting their intense attachments to place, the majority of Kazakhstan's Russians, commonly referenced in Kazakhstan today as "our Russians" (Fergana News 2020), attest to descending from several generations of settlers and "living on Kazakh land" (Kucera 2014, 4). This is in line with Kazakh imaginaries of territorial belonging, which are promoted in governmental rhetoric as tracing to pre-history. For example, a 2015 celebration positing 1465 as the state's founding-date marked 550 years of Kazakh statehood and attests to the Nur-Sultan regime's efforts to legitimize territoriality (Bondareva 2015; Lillis 2015).

Before moving on to data analysis, it is important to reassert that conceptions of *rodina* derive from a complex negotiation of multi-scalar place attachments that are themselves affected by domestic economic, political, social, and cultural contingencies, as well as, international processes and events. In the next sections, data from 2017 and 2018 surveys illustrate the relationality, elasticity, and socio-political complexity of *rodina* amongst Russians in northern Kazakhstan. Responses are juxtaposed to those of ethnic Kazakhs to demonstrate points of departure and similarity.

Research Design/Survey Administration

Two national surveys of roughly 500 respondents each were conducted in both the Russian and Kazakh languages during the spring of 2017 and autumn of 2018. These surveys were designed by the author and two colleagues as part of a national identity project. Both took place prior to President Nursultan Nazarbayev's resignation in 2019 and President Kassym-Jomart Tokayev ascendance to office. The surveys taking place in two phases has limited impact on results, as no major events occurred during the interim. The content of both surveys was nearly identical but approached different respondents.

This article analyzes only a subset of responses from the national survey. The total sample from Kazakhstan's oblasts bordering Russia (Aktobe, Atyrau, East Kazakhstan, Kostanay, North Kazakhstan, Pavlodar, West Kazakhstan), and Akmola (due to its large Russian population and proximity to the border) is 367 and includes 125 ethnic Russians and 242 ethnic Kazakhs. All respondents were citizens of Kazakhstan, age 18 or older. Surveys were administered via face-to-face interviews through representatives from Public Opinion Research Institute based in Nur-Sultan, Kazakhstan. The Kish-sampling technique was applied within specific *oblasts, raions* and settlements. While an array of options was available for the dependent variable, the analysis utilizes ethnicity so as to juxtapose Russian and Kazakh responses.

Relational Geographies of Russian-Kazakhstani *Rodina* in Northern Kazakhstan

Territorialization of Russians in northern Kazakhstan is a relational geography that includes the broader notion of a Russian World (*Russkii Mir*), Kazakhstani state territory, and the borderland region wherein seven Kazakhstani oblasts abut 12 Russian regions. The following sub-sections provide data and analysis suggesting that the term *rodina* apprehends Russian-ness as culturally broadly-defined but limitedly politicized, Kazakhstani-ness as citizenship, and locales of residence (sub-national regions, cities, and villages) in which economic, social, and cultural change is readily perceived.

SUPRA-NATIONAL TERRITORIALIZATIONS

Because the surveys conducted in 2017 and 2018 were primarily intended to ascertain degrees of civic nationalization within Kazakhstan, they did not explicitly probe significance of the *Russkii Mir* (Russian World). Nevertheless, there are questions useful in approaching this historical-civilizational imaginary as an element of place attachment for Russians in Kazakhstan's northern oblasts.

Like any form of imagined or large-scale '*rodina*,' the *Russkii Mir* may be fruitfully regarded as a mutable space that has expanded and contracted in conjunction with various contingencies over time. It is, therefore, an indefinite geographic and social imaginary that has gained traction in the post-2014 Eurasian geopolitical context. It transcends both borders and ethnicity, functioning largely as a discursive delineator between a categorically amorphous 'us' of pro-Russian-ness and a strategically malleable 'them' (O'Loughlin et al. 2017, 3; Cheskin and Kachuyevski 2019). Not only are the parameters of the term *Russkii Mir* and its manner of deployment in various post-Soviet states highly variable, but so too its reception amongst various 'compatriot' communities outside of the Russian Federation (O'Loughlin et al. 2017; Laruelle 2018; Cheskin and Kachuyevski 2019; Laruelle and Bessembayev 2019). Much of this variability relates to intensity of place attachment amongst different Russian and non-titular Russophone groups relative to their current states and smaller-scale places of residence (e.g., *raion, oblast*, borderlands).

Russian President Vladimir Putin's efforts to manage and monitor the 25-million-person Russian 'diaspora' link biopolitics to geopolitics through passportization (Artman 2013) and an appeal to spiritual linkage between *Russkii* peoples (defined by language, cultural heritage, or political ideology - Putin 2001). "Compatriots," he asserts, are those who "share a common concern for Russia's future and its people, a commitment to be useful to (the) historical homeland, to promote its socioeconomic development and strengthen its international authority and prestige" (Putin 2012). The extent to which this wide identity-net has the potential to alter the place identities of geographically concentrated people in post-Soviet borderlands is not only unclear but uneven. Table 2 demonstrates the significance of Russian language amongst the borderland population as a whole, but particularly Russians.

Russkii Mir as a supra-*rodina* takes an anti-Western defensive countenance in some outlets, and a neo-imperialist stance, a Soviet mythologization, and/or the role of cultural emollient in others. It was fear of neo-imperialism, akin to that which occurred in Ukraine, that spurred concern for northern Kazakhstan amongst pundits after 2014 (Morozov 2015; Girgas 2016). And survey data from 2017-2018 indeed points to greater concern amongst Kazakhs than Russians, as to the prospect of Russian intervention (see Table 3).

That said, rather than languishing as a "beached minority" awaiting migration or reincorporation into an historic homeland, evidence below suggests that the Russian community of northern Kazakhstan actively incorporates ideals of Kazakhstani civic nationhood and a distinctive borderlander identity into their concept of *rodina*. While appreciative of Russian transnational attention as maintaining an alternative, socio-cultural support system should economics or politics shift in Kazakhstan, Russians in the northern oblasts are highly territorialized to their small-scale places of residence and within the Kazakhstani ideal of a "common homeland" for all citizens (Nazarbayev 1992, 14).

TABLE 2—HOW IMPORTANT IS SPEAKING RUSSIAN TO YOUR IDENTITY?

ETHNICITY	NOT IMPORTANT %	SOMEWHAT IMPORTANT %	VERY IMPORTANT %
Kazakh	7	48	45
Russian	3	31	66

Data from Author Survey 2017-2018. Percentages rounded.

Table 3—I am concerned about Russian claims to Northern Kazakhstan resulting in a situation similar to Eastern Ukraine

Ethnicity	Disagree completely %	Disagree somewhat %	Agree somewhat %	Agree completely %
Kazakh	13	13	31	43
Russian	20	15	38	27

Data from Author Survey 2017-2018. Percentages rounded.

Responses to a 2020 national survey of Kazakhstan's foreign relations suggest that northern-oblast-Russians' attitudes are in line with those of the rest of the Kazakhstani population; of which 49 percent are satisfied with Kazakhstan's relationship with Russia and 24 percent would like increased ties (Laruelle et al. 2020, 3). It should be noted that 17 percent desired a lessening of the Russian presence, but only 24 percent of the national survey respondents regard Russia as a model to emulate. This suggests that the vast majority of the multi-ethnic population sanctions Kazakhstan's independent development as a state (Laruelle et al. 2020, 4).

These data in combination with the 2017-2018 survey convey northern-oblast-Russians' embrace of protections by both domestic and international law, practical networks of civil beneficence, and temporal continuity in place. The supra *rodina* of the *Russkii Mir*, therefore, has relevance to this group as Russophone peoples, but by virtue of the "frontier of friendship" concept, through which the Kazakhstani/Russian borderland has been constructed, remains nevertheless abstract and relative to more intensive, alternative scales of place attachment (Mansurov 1999). While it could become more polarized, the borderland currently functions more as a cultural emollient, enabling place attachment that is both Russian and Kazakhstani.

KAZAKHSTANI RODINA

As with any evolving modern society, social policies within Kazakhstan have transformed over the last three decades. These changes were contingent, at least in part, on demographics. The population's share of ethnic Kazakhs increased significantly due to diasporic return migration, high Kazakh birthrates, and the emigration of ethnic Russians along with many other European peoples (peaking in 1994 at 283,000 people). Over the course of the late 1990s and early 2000s, emigration stabilized to between 20,000 and 40,000 per year, until the 2006 launch of Russia's repatriation program prompted a slight uptick (Simakova 2016). In the early 2000s, emigration of ethnic Russians was partially countered by reverse migration by Russians that had departed in the 1990s (roughly 20,000 per year). This continued until 2005 when the number dropped to 10,000 and has since declined (Kazakhstan Committee on Statistics 2016). 2014 saw a revival of emigration of Russian's from Kazakhstan accounting for approximately 22,000, or 72% of all Kazakhstani emigrants between 2014 and 2017 (most of the others were Ukrainian or German – Simakova 2016). Tensions relating to the Ukraine crisis may have catalyzed some of the recent migration, but there is little evidence distinguishing economic from socio-political motives.

This demographic swing in favor of ethnic Kazakhs helped rationalize various policies reversing the Soviet process of 'russification' in favor of 'kazakhization.' These include promotion of Kazakh as the official state language, requirements for Kazakh language programing in public media, Kazakh language tests for various levels of schooling, the conversion of the alphabet from Cyrillic to Latin, and myriad public displays of Kazakh cultural heritage (e.g., place/street name changes, public iconography - Laruelle 2016a).

While some regard each of the aforementioned policies as relating to concerns over sovereignty and reflective of efforts to cement Kazakhstan's territorial integrity, no official documentation overtly confirms that conclusion. Rather, the Kazakhstani government officially couches these efforts in terms of economics, cultural diversification, and globalization of the state. For example, the Nazarbayev regime promoted the aforementioned 2015 celebrations of 550 years of Kazakh statehood as a call for all citizens to invest their futures in the state (Bondareva 2015).

> Independence was hard won by many generations of our ancestors, who defended our sacred land with blood and sweat ... Independence is the unflinching resolution of each citizen to defend Kazakhstan, their own home, and the motherland to the last drop of blood, as our heroic ancestors have bequeathed us (Nazarbayev quoted in Tengrinews 2015).

This quote and similar refrains found in the 2008 Doctrine of National Unity and the 2014 '*Mengilik El*' are representative of efforts to collectively territorialize the population across the ethnic spectrum (Rees and Burkhanov 2018; see Figure 2).

FIG. 2—My Homeland is Kazakhstan (Kazakh) and My Homeland is Kazakhstan (Russian). Source: Author Photo, Almaty.

Attention afforded the northern oblasts in this regard is encapsulated in public iconography intended to symbolize multi-ethnic unity and Kazakhstan as *rodina*. Examples include a set of statues; one pairing figures of Shokan Valikhanov (a leading Kazakh intellectual of the nineteenth century) and Russian writer Fyodor Dostoyevsky (once exiled to Kazakhstan) and another that pairs famed poets Alexander Pushkin and Abai Kunanbaev in the city of Petropavlovsk (Kucera 2014, 5-7). Such iconography is intended to convey Kazakhstan's Eurasian character.

Survey data in Table 4 supports a general perception of Kazakhstan's Eurasian-ness. Perhaps more significant, however, is Table 5's general conveyance of people's embrace of "Kazakhstani" as a moniker of place identity. While at base an ethnic toponym, it attends to a *jus soli* citizenship and territorial identity akin to "of Kazakhstan," Kazakhstanian, or Kazakhstanite. Relatedly, Tables 6, 7 and 8 point to expanding or consistent use of the term and an expectation of its sustainability. These data are indicative of at least partial fruition of a civic-national *rodina* or a 'Kazakhstani-stan' (i.e., land of the Kazakhstanis - Diener 2016).

That these results derive from people residing along the 7,000-kilometer border between Kazakhstan and Russia is, in some sense, a function of the border being at once a demarcation of sovereignty and national distinction, while simultaneously functioning as an interactive nexus (Smith 2016; Cheskin and Kachuyevski 2018, 14). Longstanding political, economic, and social relationships between state governments, regional administrations, and borderland populations denote a de jure and de facto transnational *rodina*. In other words, networks of multi-sited place attachment for many in northern Kazakhstan (both Russians and Kazakhs) include parts of Russia.

Borderland residents often have more in common with those on the other side of the border than with co-citizens throughout their state territory. It is, therefore, worth considering the degree to which a "borderland identity" is forming. Table 9's conveyance of a general consensus that it is important for citizens to regard themselves as a single political community suggests an orientation of the population toward Nur-Sultan rather than Moscow (see also Laruelle et al. 2020). In response to questions of the need for shared language and willingness for children to marry outside their ethnic community, Tables 10

TABLE 4—I WOULD CHARACTERIZE KAZAKHSTAN AS A EURASIAN COUNTRY

ETHNICITY	DISAGREE COMPLETELY %	DISAGREE SOMEWHAT %	AGREE SOMEWHAT %	AGREE COMPLETELY %
Kazakh	8	11	38	43
Russian	13	9	36	42

Data from Author Survey 2017-2018. Percentages rounded.

TABLE 5—TO WHAT EXTENT DO YOU IDENTIFY AS A KAZAKHSTANI? DO YOU IDENTIFY AS KAZAKHSTANI?

ETHNICITY	A LOT %	A LITTLE BIT %	NOT VERY MUCH %	NOT AT ALL %
Kazakh	62	17	16	5
Russian	53	23	20	4

Data from Author Survey 2017-2018. Percentages rounded.

TABLE 6—COMPARED TO 5 YEARS AGO DO YOU PERSONALLY USE THE TERM KAZAKHSTANI TO DESCRIBE YOURSELF MORE, LESS, OR ABOUT THE SAME AMOUNT?

ETHNICITY	MORE %	LESS %	ABOUT THE SAME AMOUNT %
Kazakh	42	17	41
Russian	38	16	46

Data from Author Survey 2017-2018. Percentages rounded.

TABLE 7—COMPARED TO 5 YEARS AGO DO YOU HEAR THE TERM KAZAKHSTANI BEING USED BY OTHERS TO DESCRIBE THEMSELVES MORE, LESS, OR ABOUT THE SAME AMOUNT?

ETHNICITY	MORE %	LESS %	ABOUT THE SAME AMOUNT %
Kazakh	51	13	36
Russian	37	12	51

Data from Author Survey 2017-2018. Percentages rounded.

TABLE 8—FUTURE GENERATIONS, REGARDLESS OF ETHNICITY, WILL CALL THEMSELVES KAZAKHSTANI

ETHNICITY	DISAGREE COMPLETELY %	DISAGREE SOMEWHAT %	AGREE SOMEWHAT %	AGREE COMPLETELY %
Kazakh	3	6	26	65
Russian	7	6	36	51

Data from Author Survey 2017-2018. Percentages rounded.

and 11 suggest both Russians and Kazakhs possess similar attitudes pertaining to these issues.[5] While Tables 5, 6, 7, and 8 (above) suggest that Kazakhs and Russians in the northern oblasts share a certain place identity (Kazakhstani) and tend to regard the state as requiring a unified political community (Table 9), ethnic identity remains pertinent in the realm of language and interpersonal relationships for certain segments of the population (Tables 10 and 11).

TABLE 9—IT IS IMPORTANT FOR ALL THE ETHNIC GROUPS THAT MAKE UP THE CITIZENS OF KAZAKHSTAN TO CONSIDER THEMSELVES UNITED AS A SINGLE POLITICAL COMMUNITY

ETHNICITY	DISAGREE COMPLETELY %	DISAGREE SOMEWHAT %	AGREE SOMEWHAT %	AGREE COMPLETELY %
Kazakh	2	3	27	68
Russian	7	3	35	55

Data from Author Survey 2017-2018. Percentages rounded.

TABLE 10—IN YOUR OPINION, HOW IMPORTANT IS IT TO HAVE A SINGLE SHARED LANGUAGE IN A COUNTRY?

ETHNICITY	NOT IMPORTANT %	SOMEWHAT IMPORTANT %	VERY IMPORTANT %
Kazakh	36	49	15
Russian	22	51	27

Data from Author Survey 2017-2018. Percentages rounded.

TABLE 11—I WOULD APPROVE OF MY CHILDREN MARRYING SOMEONE OF A DIFFERENT ETHNIC GROUP

ETHNICITY	DISAGREE COMPLETELY %	DISAGREE SOMEWHAT %	AGREE SOMEWHAT %	AGREE COMPLETELY %
Kazakh	22	16	26	36
Russian	14	10	28	48

Data from Author Survey 2017-2018. Percentages rounded.

Table 12's overwhelming assertion of capacities for patriotism even without the embrace of the Kazakhstani moniker may be reflective of borderland-place-attachment that is relational, thus constituting a "both/and" place-identity rather than an "either/or" (Antonsich 2009). Membership in state's citizenry is therefore not a matter of "where you live geographically, (but) your mentality, your aspirations and … the person's self-identification" (Putin 2001). That a vast majority of northern-oblast-Russians share perceptions with their ethnic

TABLE 12—I CAN BE PATRIOTIC AND FEEL THAT I BELONG IN KAZAKHSTAN WITHOUT CALLING MYSELF KAZAKHSTANI

ETHNICITY	DISAGREE COMPLETELY %	DISAGREE SOMEWHAT %	AGREE SOMEWHAT %	AGREE COMPLETELY %
Kazakh	1	9	28	62
Russian	1	7	40	43

Data from Author Survey 2017-2018. Percentages rounded.

TABLE 13—I FEEL THAT I BELONG AND AM FULLY WELCOME TO LIVE AND FLOURISH IN KAZAKHSTAN

ETHNICITY	DISAGREE COMPLETELY %	DISAGREE SOMEWHAT %	AGREE SOMEWHAT %	AGREE COMPLETELY %
Kazakh	0	6	25	69
Russian	4	3	26	67

Data from Author Survey 2017-2018. Percentages rounded.

Kazakh neighbors of being welcome and capable of flourishing in Kazakhstan is a strong statement as to collective territorialization (see Table 13).

The above data affords ample evidence that a civic national identity territorialized within the boundaries of Kazakhstan has emerged amongst both Russians and Kazakhs of the northern oblasts. While some people adhere to ethnic conceptions of self and group that relate to alternative territorializations (i.e., alt-*rodini* – Rees et al. forthcoming), these do not seem to offset the broader ideals of civic belonging cultivated over the last 30 years. The degree to which these identities may coexist or conflict has much to do with their personal, localized, place attachments, as well as, material and economic conditions, perceptions of parity in civil beneficence, and ethnic relations in the post-Nazarbayev era (Jašina-Schäfer's 2019). The following section explores this intra-state regional scale of place attachment.

INTRA-STATE REGIONAL TERRITORIALIZATIONS

Like any large state, Kazakhstan is an agglomeration of distinct regions possessing unique cultural and social attributes. A north/south dichotomy within Kazakhstan is particularly prevalent in public discourse. Kazakh nationalists and southerners frequently portray the north as urban, "modern, culturally and linguistically Russified, and civilized" (Koch 2014, 438). By contrast, the south carries a stigma of being rural and "coded as traditional, uncivilized, and culturally and linguistically 'Kazakh'" (Koch 2014, 438). Such divisions are tied up with class, race, and interactions with neighboring states (Koch and White 2016).

Cummings (2006, 149) suggests that ethnic Kazakhs in the northern oblasts view Russia as non-threatening. This relates, in part, to cross-border social and economic networks. The Russia/Kazakhstan border contrasts the Chinese/Kazakhstan border in its lack of consistent demarcation and its traversal by integrated transport (i.e., 200 major roadways, 16 rail lines), electricity, industrial infrastructure, and active social networks (Mansurov 1999, 371-380). A 2020 survey suggests that 55 percent of the Kazakhstan's population desire a lesser Chinese presence in Kazakhstan with only 24 percent satisfied with the PRC's current role (Laruelle et al. 2020, 5). Viewed through a relational geographic lens, despite large-scale, bilateral trade-relations with Kazakhstan, China represents an "other" against which much of the population defines itself. By contrast,

'Russian-ness' in the north (as well as Almaty) is seemingly part of a broader concept of Kazakhstan (Koch 2014; Cheskin and Kachuyevski 2019).

This point is made acutely in a June 3, 2020 interview wherein President Kassym Zhomart Tokaev actually forbade the use of the word "diaspora" in reference to Russians in Kazakhstan (Fergana News 2020). Tokaev stated, "more than 3.5 million Russians live in Kazakhstan, their share in the total population is almost 19 percent. The Russians living in our country are an integral part of our people, they enjoy all rights." He also emphasized that Kazakhstanis, regardless of their ethnic origin, are a single people, and suggested that the concept of "national minority" had no place in the republic.

Rather ironically, akin to Russia's repatriation program, the Nur-Sultan regime has periodically encouraged the ethnic Kazakh "diaspora" to 'return' to Kazakhstan. Many of these *Oralmandar* (ethnic Kazakh returnees) have been directed through state programs to settle in the northern oblasts (Diener 2008). While it is impossible to unequivocally attribute such settlement to state interests in diluting Russian populations, rumors of an April 2014 plan to shift some 300,000 ethnic Kazakhs (most *Oralmandar*) to the north seemed linked to the Ukraine crisis (Goble 2014). Whatever the motive, settlement of Kazakh 'return migrants' presents localized social complexities for territorialization amongst both migrants and receiving populations.

Many Kazakh *Oralmandar* from outside former Soviet Republics have limited Russian language skills, which makes adaptation to the culturally Russified north difficult (Diener 2008). Relatedly, the introduction of people regarded by both Russian and Kazakh residents of the northern oblasts as 'foreigners' disrupts the status quo and evokes questions of equitable civil beneficence (education, pensions, housing, etc. – see Kusçu 2014; Werner et al. 2017). Kazakhstan's repatriation program was terminated in 2012 but then resumed with a March 2014 official order that all returnees must settle in Akmola, Atryau, West Kazakhstan, Kostanay, Pavlodar, North-Kazakhstan, and East-Kazakhstan provinces (Mamashuli 2014). All of these provinces are regarded as part of the Kazakhstan/ Russian borderland. According to an August 3, 2014 RFERL Report, the order for targeted settlement was rescinded soon after its issue and *Oralmandar* may now settle in any of Kazakhstan's fourteen provinces (Panier 2014).

In 1989, 66% of Russian residents in the Kazakh SSR were born there, while 37% of those born elsewhere had lived in the republic for over 20 years (Peyrouse 2007). One may therefore infer that the ethnic 'un-mixing' of the 1990s and more recent departures have left either those lacking the resources to migrate or with the most ardent claims to territorial belonging. Tables 14 and 15 suggest parallels in confidence in personal futures and the future of one's children within Kazakhstan.

While Table 16 attests to a majority of respondents regarding ethnic relations as peaceful and stable, an influx of ethnic Kazakhs in the north has the potential to

Table 14—I am confident in my future in Kazakhstan

ETHNICITY	DISAGREE COMPLETELY %	DISAGREE SOMEWHAT %	AGREE SOMEWHAT %	AGREE COMPLETELY %
Kazakh	1	7	28	64
Russian	8	5	32	55

Data from Author Survey 2017-2018. Percentages rounded.

Table 15—I am confident in the future of my children in Kazakhstan

ETHNICITY	DISAGREE COMPLETELY %	AGREE SOMEWHAT %	AGREE COMPLETELY %
Kazakh	3	42	57
Russian	10	38	45

Data from Author Survey 2017-2018. Percentages rounded.

alter perceptions of *rodina* by enhancing the Kazakh population and de-emphasizing Russian language and traditions. This is particularly true since the decline in Kazakhstan's Russian population from its peak (40%) in 1979 to today (19%) will likely continue as a result of natural aging and low birth rates. The natural growth rate for Kazakhs in 2015 was 2.3% and -0.51 for Russians. With the average age of Kazakhs at 28.9 and 38.5 for Russians in 2013, a pending generation gap will render the north increasingly Kazakh through natural means and, in 20-30 years, see a pronounced expansion of Russian pensioners (Laruelle 2018, 71).

Building upon the 'graying' of the Russian population is the fact that the northern oblasts have demonstrated declining economic growth. This is especially so in Akmola, Kostanay, Pavlodar, North Kazakhstan, and East Kazakhstan oblasts (see Figure 3). While Almaty and Nur-Sultan dominate the state's economic geography, wealth also concentrates in the two west Caspian oblasts of Atyrau and Mangystau and in the agricultural sector of South Kazakhstan and the Skymkent region. The northern oblasts host a number of atrophied industries and have seen recent national economic policies shift development efforts from a broad national

Table 16—In general, ethnic relations in Kazakhstan are peaceful and stable

ETHNICITY	DISAGREE COMPLETELY %	DISAGREE SOMEWHAT %	AGREE SOMEWHAT %	AGREE COMPLETELY %
Kazakh	5	4.5	23	72
Russian	2	3	26	69

Data from Author Survey 2017-2018. Percentages rounded.

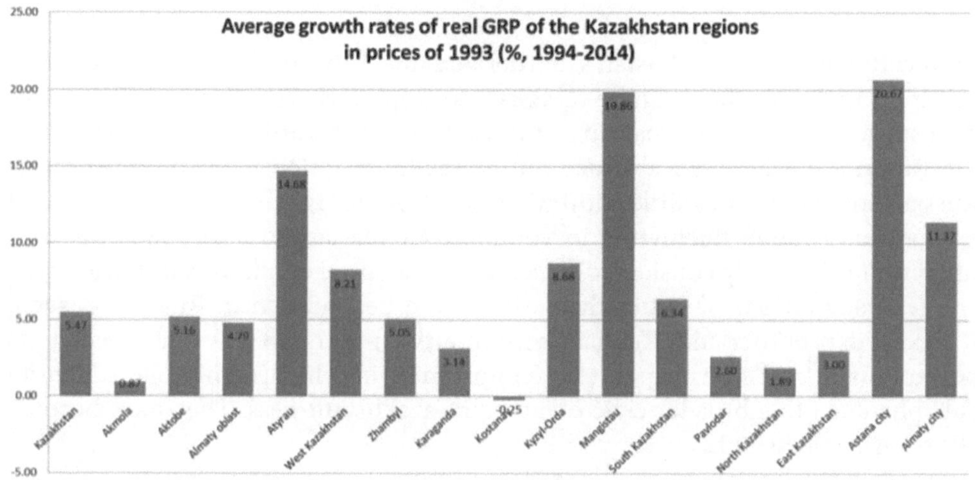

FIG. 3—Average Growth Rates in Real GRP for Kazakhstan's Regions in Prices of 1993 (1994-2014). Source: Kazakhstan National Statistics Agency 2015.

distribution to concentrations within specific regions and cities (Turgenbayev and Diener 2018). This a has the potential to negatively affect place attachment is rural settlements and many of the single industry cities of the north.

While Kazakhstan's economic success over the last two decades has benefited the spectrum of ethnic groups within the state (Ó Beacháin and Kevlihan 2013, 346), there is a distinct disparity in its regional economic geography (Turgenbayev and Diener 2018). Such disparity brings into question the distribution of civil beneficence, which can problematize conceptions of place attachment. Jašina-Schäfer's (2019, 15-16) interviews with Russophone youth in Petropavlovsk and Nur-Sultan reveal their consideration of economics and stage of life contingencies in regard to *rodina* conceptions. The colloquial phrase "(r)*odina* is where your bottom feels good" speaks directly to this point. It is through a multi-scalar relational perception of "comfort" that the majority of Russians demure an imminent abandonment of current residence (see Table 17).

TABLE 17—I HAVE NO PLANS TO MOVE OUT OF MY *RODINA*

ETHNICITY	DISAGREE COMPLETELY %	DISAGREE SOMEWHAT %	AGREE SOMEWHAT %	AGREE COMPLETELY %
Kazakh	3	6	21	70
Russian	4	7	30	59

Data from Author Survey 2017-2018. Percentages rounded.

Conclusion

Ethnic Russians of Kazakhstan's borderland the Russian Federation have had nearly 30 years to either migrate or agitate for separatism/irredentism. The above data suggest that the vast majority of those Russians regard their membership in the Russian World as an abstract cultural support system and the prospect of Russian intervention as little more than a safeguard against Kazakh nationalist extremism. Though fluctuating in volume over the last 30 years, the option of migration is limitedly considered as economic or social conditions in Kazakhstan are generally on par or better than what could be expected in Russia. As such, Kazakhstan's borderland Russian community appears to have negotiated its perception of *rodina* as a point of compromise in which ethnic place identity is hyphenated to a broader civic or *jus soli* Kazakhstani-ness. They have become "Russian-Kazakhstanis."

Essential to this process is the elasticity of their conceptions of *rodina*. As this article relays, the term represents a perspective on socio-spatial belonging that often eludes both scholarship and policy by effectively apprehending a mutable and multi-scalar modality of territorialization. Amongst many Russians outside the Russian Federation, this problematizes the primordialistic/*jus sanguinis* ideals of identity in favor of constructivist/*jus soli* notion attending to places of residence. Geopolitical, economic, and social contingencies are, nevertheless, pertinent factors in an ongoing negotiation of *rodina*.

As anticipated within relational geographies of place attachment, perceptions of belonging and alienation manifest in conjunction with contingencies of varied sites and scales of place in an ongoing process of de- and re-territorialization. The process lends to a more idiographic than nomothetic approach to the phenomenon, which is, at its core, an emotional commitment to place. This research suggests that a more elastic conceptualization of place attachment affords better understanding of the mutability of place-identities and their capacity to be both multi-sited and varying in scale.

While the term *rodina* may well be idiomatic, it nevertheless succeeds in capturing both the nebulous notion of geographic belonging and mutable aspect of place attachment absent from many studies. Through a relational approach to territorialization, we can better understand its capacity to constrict and localize but also expand to incorporate changing ideals and components of imagined communities. Deepening our understanding of this process is a key step in addressing catalysts of conflict and managing adaptations to dynamic fields of corporeal mobility, instantaneous communication technology, and integrated economic systems.

Funding

This work was supported by the Association of American Geographers [Summer Research Scholarship]; University of Kansas [General Research Fund].

Notes

1. Khrushchev's 1954 'gifting' of Crimea to Ukraine has been portrayed as an historical wrong to be righted within Russia (Coynash and Charron 2019). Moreover, the Soviet delineation of the Kazakh ASSR/SSR and Russia border occurred in stages of 'exchange' between 1930 and 1965 with a final border agreement registered in 2005.

2. Sevastopol holds special meaning for Russians since the Crimean War (1853-1856). It's role as a base for the Black Sea Fleet also plays into Russian place attachment (Biersack and O'Lear 2014; Charron 2016).

3. A natural/civilizational divide between the Russian Empire and the "West" was asserted by the Polish-Lithuanian Commonwealth (Snyder 1998).

4. Autochthonous is a problematic term in place attachment research. Though beyond the scope of this paper, it is a relevant topic for future research.

5. For comparison across time see 1998 survey by Barrington (2001, 138).

References

Agnew, J. 2009. *Globalization and Sovereignty.* Lanham: Rowman and Littlefield.

Anthias, F. 2002. Where Do I Belong? Narrating Collective Identity and Translocational Positionality. *Ethnicities* 2 (4): 491–514.

Antonsich, M. 2009. On Territory, the Nation-State, and the Crisis of the Hyphen *Progress in Human Geography* 33(6): 789–806.

_____ 2010. Searching for Belonging: An Analytical Framework. *Geography Compass* 4 (6): 644–659.

Artman, V. 2013. Documenting Territory: Passportization, Territory, and Exception in Abkhazia and South Ossetia. *Geopolitics* 18 (3): 682–704.

Barcus, H. and S. Brunn. 2010. Place Elasticity: Exploring a New Conceptualization of Mobility and Place Attachment in Rural America. *Geografiska Annaler: Series B, Human Geography.*

Barrington, L. 2001. Russian-Speakers in Ukraine and Kazakhstan: "Nationality," "Population," or Neither? *Post-Soviet Affairs,* 17 (2): 129–158.

Biersack, J. and S. O'Lear. 2014. The Geopolitics of Russia's Annexation of Crimea: Narratives, Identity, silences, and Energy. *Eurasian Geography and Economics,* 55 (3): 247–269.

Bondareva. V. 2015. *Semki serial o Kazakhskom Hanstve Nachnutsya Yanvarye 2015 Roda.* B News KZ [http://bnews.kz/ru/news/post/237306/].

Brubaker, R. 2011. Nationalizing States Revisited: Projects and Processes of Nationalization in Post-Soviet States. *Ethnic and Racial Studies,* 34 (11): 1785–1814.

Charron, A. 2016.Whose is Crimea? Contested Sovereignty and Regional Identity. *Region: Regional Studies of Russia, Eastern Europe, and Central Asia* 5 (2): 225–256.

Cheskin, A. and A. Kachuyevski. 2019. The Russian Speaking Populations in the Post-Soviet Space: Language, Politics, and Identity. *Europe Asia Studies* 71(1): 1–23.

Chinn, J. and R. Kaiser 1996. *Russians as the New Minority.* Westview Press.

Coynash, H. and A. Charron, 2019. Russian-Occupied Crimea and the State of Exception: Repression, Persecution, and Human Rights Violations. *Eurasian Geography and Economics* 60 (1): 28-53.

Cummings, S. 2006. Legitimation and Identification in Kazakhstan. *Nationalism and Ethnic Politics* 12 (2): 177–204.

Diener, A.C. 2008. Settlement of the Returning Kazakh Diaspora: Practicality, Choice, and the Nationalization of Social Space. In *Homeland, Migration, and Belonging in Eurasia,* Washington, DC: Woodrow , edited by C. Buckley and B. Ruble (eds.). Wilson Press. 265–304.

_____ 2015. Assessing Potential Russian Irredentism and Separatism in Kazakhstan's Northern Oblasts. *Eurasian Geography and Economics* 56 (5): 469–492.

_____ 2016. Imagining Kazakhstani-stan: Negotiations of Homeland and Titular-Nationality. In *Kazakhstan in the Making: Legitimacy, Symbols, and Social Changes,* edited by M. Laruelle, 131–154. Lanham, Lexington Books.

Doel, M. 2007. Posts Structuralist Geography: A Guide to Relational Space by J. Murdoch. *Annals of the Association of American Geographers* 97 (4): 809–810.

Fergana News 2020. Tokaev Zapretul Nazivat Russkikh in Kazakhstanye "Diaspora" (Tokaev Forbids Reference to Russians in Kazakhstan as "Diaspora"). *Fergana News*, 3 June 2020 [https://fergana.news/news/118757/].

Girgas, A. 2016. *Beyond Crimea: The New Russian Empire.* New Haven: Yale University Press

Goble, P. 2014. Astana Shifting Ethnic Kazakhs to Northern Kazakhstan to Block Any Russian Threat. *The Interpreter.* [http://www.interpretermag.com/astana-shifting-ethnic-kazakhs-to-northern-kazakhstan-to-block-any-russian-threat/].

Hediger, R. 2019. *Homesickness: Of Trauma and the Longing for Place in a Changing Environment.* Minneapolis: University of Minnesota Press.

Herb, G. and D. Kaplan (eds.) 2018. *Scaling Identity.* Lahman: Rowman and Littlefield Ingram, A. 2001. Broadening Russia's Borders? The Nationalist Challenge of the Congress of Russian Communities. *Political Geography* 20 (2): 197–208.

Jašina-Schäfer, A. 2019. Where Do I Belong? Narratives of Rodina among Russian-Speaking Youth in Kazakhstan. *Europe-Asia Studies* 71 (1): 97–116.

Jones, M. 2009. Phase Space: Geography, Relational Thinking, and Beyond. *Progress in Human Geography* 33 (4): 487–506.

Kaiser, R. and E. Nikiforova 2006. Borderland Spaces of Identification and Dis/location: Multiscalar Narratives and Enactments of Seto Identity and Place in the Estonian-Russian Borderlands. *Ethnic and Racial Studies* 29 (5): 928–958.

Kazakhstan Committee on Statistics of the Ministry of National Economy. 2016. *Statistical Yearbook of Demographic and Economic Data 1999-2015.* Almaty KZ.

Kazakhstan National Agency for Statistics 2015. *Valovoi Regional'nyi Produckt Respubliki Kazakhstan Za* 2014.

Kirkness, P. and A. Tijé-Dra (eds.) 2017. *Negative Neighborhood Reputation and Place Attachment: The Production and Contestation of Territorial Stigma.* London: Routledge.

Koch, N. 2014. Bordering on the modern: Power, Practice, and Exclusion in Astana. *Transactions of the Institute of British Geographers* 39 (3): 432–443.

Koch, N. and K. White. 2016. Cowboys, Gangsters, and Rural Bumpkins: Constructing the 'Other' in Kazakhstan's 'Texas'. In *Kazakhstan in the Making: Legitimacy, Symbols, and Social Changes*, edited by M. Laruelle, 181–208.Lanham: Lexington Books.

Kucera, J. 2014. Northern Kazakhstan isn't the Next Crimea – Yet *Aljazeera America.* [http://america.aljazeera.com/articles/2014/6/19/north-kazakhstanisntthenextcrimeaayet.html].

Kusçu, I. 2014. Ethnic Return Migration and Public Debate: The Case of Kazakhstan. *International Migration* 52 (2): 178–197.

Laruelle, M. 2015. Kazakhstan's Posture in the Eurasian Union: In Search of Serene Sovereignty. *Russian Analytical Digest* 165: 7–10.

_____ 2016a. Which Future for National Patriots? The Landscape of Kazakh Nationalism, In *Kazakhstan in the Making: Legitimacy, Symbols, and Social Changes*, edited by M. Laruelle, 155–180. Lanham, Lexington Books.

_____ 2016b. The Three Colors of Novorossiya, or the Russian Nationalist Mythmaking of the Ukrainian Crisis. *Post Soviet Affairs* 32 (1): 55–74.

_____ 2018. Why No Kazakh Novorossiya? Kazakhstan's Russian Minority in a Post-Crimea World. *Problems of Post-Communism* 65 (1): 65–78.

Laruelle, M., D. Royce, and S. Beyssembayev, 2019. Untangling the Puzzle of "Russia's Influence" in Kazakhstan. *Eurasian Geography and Economics* 6 (2): 211–243.

Laruelle, M., G. Toal, J. O'Loughlin, and K. Bakke, 2020. Kazakhs are Wary of Neighbours Bearing Gifts. *Open Democracy.* [https://www.opendemocracy.net/en/odr/kazakhs-are-wary-neighbours-bearing-gifts/].

Lewicka, M. 2011. Place Attachment: How Far Have We Come in the Last 40 Years? *Journal of Environmental Psychology* 31 (3): 207–230.

Lillis, J. 2015. 'Kazakhstan Celebrates Statehood in Riposte to Russia. *Eurasianet.* [http://www.eurasianet.org/node/71536].

Malpas, J. 2018. *Place and Experience: A Philosophical Topography.* Cambridge: Cambridge University Press.

Mamashuli, A. 2014. 'Opredeleni Sem Oblastei Dlya Rasseleniya Oralmanov' (Seven Areas for Resettlement of Oralmans identified) RFERL. [https://rus.azattyq.org/a/rasseleniye-oralmanov-po-oblastyam-postanovleniye/25318268.html].

Mansurov, T. 1999 Kazakhstansko-Russiisco Otnosheniya – Stratetriya Vechoi Druzhbii (Kazakhstani-Russian Relationship – Strategy for Eternal Friendship) in *Ambassadorial Years in Moscow 1994-2002* Moscow: Real Press.

Massey, D. 1993. Power -Geometry and Progressive Sense of Place. In *Mapping the Futures: Local Cultures, Global Change*, edited by J. Bird et al. (eds.), 60-70. London: Routledge.

_____ 2006. Landscape as Provocation – Reflections on Moving Mountains. *Journal of Material Culture* 11 (1–2): 33–48.

Megoran, N. 2017. *Nationalism in Central Asia: A Biography of the Uzbekistan-Kyrgyzstan Border*. Pittsburg: University of Pittsburg Press.

Morozov, V. 2015. Kazakhstan and the 'Russian World' Is Intervention on the Horizon? *PONARS Eurasia Policy Memo* 364: 1–6.

Murdoch, J. 2006. *Post Structuralist Geography: A Guide to Relational Space*. London: Sage.

Nazarbayev, N. 1992. *A Strategy for the Development of Kazakhstan as a Sovereign State*. Almaty: Ministry of Foreign Affairs.

O'Loughlin, J., G. Toal and V. Kolossov 2017. Who Identifies with the "Russian World"? Geopolitical Attitudes in Southeastern Ukraine, Crimea, Abkhazia, South Ossetia, and Transnistria. *Eurasian Geography and Economics* 57 (6): 745–778.

Ó Beacháin, D. and R. Kevlihan. 2013. Threading the Needle: Kazakhstan between Civic and Ethno-Nationalist State Building. *Nations and Nationalism* 19 (3): 337–356.

Paasi. A. 1995. Constructing Territories, Boundaries, and Regional Identities. In *Contested Territory: Border Disputes at eh Edge of the Former Soviet Empire* edited by T. Forsberg. Aldershot: Edward Elgar Publishing.

Pannier, B. 2014. 'A Tale of Russian Separatism in Kazakhstan' *RFERL*. [http://www.rferl.org/content/qishloq-ovozi-kazakhstan-russian-separatism/25479571.html].

Peyrouse, S. 2007. Nationhood and the Minority Question in Central Asia: The Russians in Kazakhstan. *Europe Asia Studies* 59 (3): 482–501.

Ponzetti, J.J. 2003, Growing Old in Rural Communities: A Visual Methodology for Studying Place Attachment. *Journal of Rural Community Psychology* 6 (1): 1–11.

Putin, V. 2001. Vystuplenie Presidenta Rossiiskoi Federatsii V. V. Putina na Kongresse sootechestvennikov (Speech of the Russian President V.V. Putin at the Congress of Compatriots) October 11. [http://old.nasledie.ru/politvnt/19_44/article.php?art=24].

_____ 2012. Vladimir Putin's Address to the Participants of the Fourth World Conference of Compatriots, October 26. [http://en.kremlin.ru/events/president/news/16719].

Rees, K. M. and Burkhanov, A. 2018. Constituting the Kazakhstani Nation: Rhetorical Transformation of National Belonging. *Nationalism and Ethnic Politics* 24 (4): 433–455.

Rees, K. M. and N. Webb Williams 2017. Explaining Kazakhstani, Identity: Supraethnic, Ethnicity, Identity, Language, and Citizenship. *Nationalities Papers* 45 (5): 815–839.

Rees, K.M., N. Webb-Williams, and A. C. Diener. Forthcoming. Territorial Belonging and Homeland Disjuncture: Uneven Territorializations in Kazakhstan. *Europe Asia Studies*.

Riabchuk, M. 2015. Two Ukraines Reconsidered: The End of Ukrainian Ambivalence? *Studies in Ethnicity and Nationalism* 15 (1): 138–156.

Rogers, A., Castree, N., and Kitchin, R. (eds). 2013. *Oxford Dictionary of Human Geography*. Oxford: Oxford University Press.

Schiller, N. G. 2005. Transnational Social Fields and Imperialism: Bringing a Theory of Power to Transnational Studies. *Anthropological Theory* 5(4): 439–461.

Shevel, O. T 2019. The Politics of Dual Citizenship in Post-Soviet States. *PONARS Eurasia Policy Memo* 587, 1–6.

Simakova 2016. *Emigratsiiya Russkikh Preglashenie k Razmyshleniiu* (Russian Eemigration: An Invitation for Reflection). *Kazakhstan Spektr* 75 (1): 99-109.

Smith, J. 2016. The Transformation of Soviet Republic Borders to International Borders: Competing Concepts of the Kazakhstan Russia Border. *Journal of Borderland Studies* 23 (1): 91–104.

Snyder T. 1998. The Polish-Lithuanian Commonwealth Since 1989: National Narratives in Relations among Poland, Lithuania, Belarus and Ukraine. *Nationalism and Ethnic Politics*, 4 (3): 1–32.

Tengrinews 2015. *Kazakhskoi Gosudarstvennosti v 2015 Rodu Ispolnitsya 550 Let – Nazarbayev, TengriNews.* [http://tengrinews.kz/kazakhstan_news/kazahskoy-gosudarstvennosti-2015-godu-ispolnitsya-550-let-263876/].

Turgenbayev, Y. and A. C. Diener 2018. Kazakhstan's Evolving Regional Economic Policy: Assessing Strategies of Post-Socialist Economic Development. *Eurasian Geography and Economics* 59 (5–6): 657–684.

Trimbach, D.J. and S. O'Lear. 2015. Russians in Estonia: Is Narva the next Crimea? *Eurasian Geography and Economics* 56 (5): 493–504.

Verdery, K. 1999. *The Political Lives of Dead Bodies.* New York: Columbia University Press.

Werner, C., Emmelhainz, C., and Barcus, H. 2017 'Privileged Exclusion in Post-Soviet Kazakhstan: Ethnic Return Migration, Citizenship, and the Politics of (Not) Belonging', *Europe-Asia Studies* 69 (10): 1557–1583.

Zhirinovsky, V. 2014. Zhirinovsky's 23 February 2014 Declaration. [https://www.youtube.com/watch?v=EpMj8_1Iz_E].

PLACES OF MEMORY, HISTORIC PRESERVATION, AND PLACE ATTACHMENT IN NAZI GERMANY

JOSHUA HAGEN

ABSTRACT. The Nazi Party launched a series of ambitious building programs soon after seizing power intended to achieve a comprehensive spatial reorganization of Germany. As part of these expansive building programs, the regime invested considerable time, effort, and resources in creating an archipelago of places of memory substantiating Nazi ideology and power. This commemorative topography stitched together a profusion of places and spaces into a totalitarian liturgy that glorified notions of struggle and sacrifice on behalf of Hitler and his movement. This article investigates how this topography of Nazi memory worked to translate intangible ideologies into tangible places and experiences, and in doing so, highlights the mutable and multifaceted relationships between places of memory, historic preservation, and place attachment. *Keywords: historic preservation, nationalism, Nazi Germany, place attachment, places of memory.*

After seizing power in 1933, the National Socialist regime and its Führer Adolf Hitler launched a series of ambitious building programs across Germany intended to achieve a comprehensive reorganization of the country's civic, residential, socioeconomic, and cultural geographies. The planning and construction of massive government centers, rally grounds, residential districts, and the autobahn have received considerable attention, for example, as have the regime's military-industrial-concentration camp complexes (Jaskot 2000; Spotts 2003; Giaccaria and Minca 2016; Hagen and Ostergren 2020). It is often overlooked that the regime placed considerable importance on creating an archipelago of places of memory substantiating Nazi ideology and power. This commemorative topography stitched together a profusion of places and spaces into a totalitarian liturgy that glorified notions of struggle and sacrifice on behalf of Hitler and his movement. Like any liturgy, this Nazi narrative featured its own timeline punctuated by seminal tragedies and triumphs grounded in quasi-sacred places of memory. This article examines how this topography of Nazi memory worked to translate intangible ideologies into tangible places and experiences, and in doing so, highlights the mutable and multifaceted relationships between places of memory, historic preservation, and place attachment in conditioning specific mentalities and behaviors. Specifically, the Nazi movement sought to shape places of memory that grounded its relatively short and murky history within a larger spatial narrative of racial struggle stretching from prehistory to a promised glorious future for the German nation under Hitler's thousand-year Reich.

Nazi officials of various ranks recognized the centrality of collective memory in conditioning the masses into unquestionable belief and obedience, and the importance of place and place attachment in solidifying and sharing memory. In anatomy, memory is understood as an aggregation of electro-chemical signals exchanged across networks of neurons and synapses concentrated in the brain. In psychology, memory is studied as a series of interrelated cognitive processes involving the acquisition, interpretation, storage, and retrieval of personal experience and accumulated knowledge that helps individuals make sense of the past, act in the present, and plan for the future. Memory is also intimately tied to sociology as humans are inherently social beings. Commonly credited to sociologist Maurice Halbwachs, the notion that individuals situate, interpret, store, and retrieve memories within the context of social groups and settings is well accepted across the social sciences and humanities (Connerton 1989; Halbwachs 1992; Erll 2005). As noted by cultural theorist Jan Assmann:

> The past exists, if it can be said to exist at all, in a double form: as a sedimentation of relics, traces, and personal memories and as a social construction. This dual nature characterizes the personal past that is with us human beings not only as internal memory traces and external memory symbols of every sort but also as an image or narrative that we construe and carry with us as our autobiographical or episodic memory. (2011, 15)

From that dual perspective, the social characteristics of memory help undergird frameworks for group formation, belonging, action, and cohesion, as well as their opposites.

In addition to memory being inherently social, geographers and social theorists similarly regard humans as inherently spatial, as continually engaged in the production of space and place. These perspectives view human spatiality as a set of dynamic and deliberate strategies to establish control over nature, social relations, and meaning. If humans are inherently social and spatial beings, it is hardly surprising that memory, both individual and collective, is grounded in space and place (Nora 1989; Assmann 2009; De Nardi and others 2020). The intrinsic spatiality of memory also encompasses a set of dynamic and deliberate strategies to give tangible form to ephemeral experiences. Humans employ the social production of space and place as a means to produce, share, and perpetuate notions of collective memory, or a set of shared understandings of the past and its meaning for the present and future held intrinsic to group identity.

The processes of "making places" are, therefore, intrinsically entangled with the processes of "making memories," among other facets drawing together human spatiality and social relations (Hoelscher and Alderman 2004; Foote and Azaryahu 2007; Meusburger, Heffernan, and Wunder 2011; Jones and Garde-Hansen 2012; Maus 2015). As geographer Michael Strong explained: "Particularly poignant encounters with place become embedded within the spaces of the brain's neural networks available for recall when the appropriate connotative signals draw them from the recesses of memory into the here-and-now of experience" (2018, 20). The strategies and practices associated with historical preservation provide one means to foster those "particularly poignant encounters with place" and thereby promulgate collective memories and place attachments.

Historic Preservation, Place Attachment, and National Identity

The spatiality of collective memory and historic preservation encompasses several interrelated and iterative processes beginning with the selection of people, events, and narratives worthy of commemoration and associating them with representative places. Those processes of selection invariably engender efforts to stage those places of memory, ranging from simple markers to comprehensive reconstructions, and efforts to promote the consumption of places of memory, ranging from casual visits to extravagant performances. Finally, these processes are always loaded with values and ideologies. As geographers Derek H. Alderman and Joshua F. J. Inwood noted, "while memory is ostensibly about the past, it is shaped to serve ideological interests in the present and to carry certain cultural beliefs into the future" (2013, 187). Taken together, the varied processes of selection, staging, and consumption associated with historic preservation contribute to the social production of places of memory and thereby cultivate notions of place attachment (Lewicka 2014).

Place attachment is commonly defined as the "symbolic relationship formed by people giving culturally shared emotional/affective meanings to a particular space or piece of land that provides the basis for the individual's and group's understanding of and relation to the environment. … Thus, place attachment is more than an emotional and cognitive experience, and includes cultural beliefs and practices that link people to place," according to anthropologist Setha M. Low (1992, 165). In that vein, psychologists Leila Scannell and Robert Gifford conceptualized place attachment through a tripartite framework of person, place, and process (2010, 2). Environmental psychologists, sociologists, and urban and environmental planners, among other scholars, have devoted considerable time to investigating the relationships between various demographic characteristics and intensity of place attachment, i.e. person.

In contrast, place—encompassing its physical and social dimensions—and process—encompassing affect, cognition, and behavior—have garnered considerably less scrutiny (Lewicka 2011). Historic preservation provides one avenue to illuminate the nexus of person, place, and process that conditions and binds together the social, physical, and ideological dimensions of place attachment. Historic preservation comprises a set of socially mediated practices intended to manipulate the social and physical dimensions of place to communicate shared memories of the past and thereby influence actions in the present and expectations for the future. In simpler terms, historic preservation can be understood as the processes through which people create places of collective memory, including selection, staging, and consumption, as noted previously. By extension, historical preservation helps bind together place and memory though "particularly poignant encounters" (Strong 2018, 20) that confer "culturally shared emotional/affective meanings to a particular space or piece of land" (Low 1992, 165) and thereby engender collective memories and place attachments.

The social production of places of memory and place attachment can buttress senses of belonging and ideology across multiple scales and registers, but backed by the official sanction of the nation-state, notions of national identity and homeland gradually assumed predominate roles since the nineteenth century. Political scientist Duncan S. A. Bell's formulation of a national mythscape, or a "discursive realm, constituted by and through temporal and spatial dimensions, in which the myths of the nation are forged, transmitted, reconstructed and negotiated constantly," can help conceptualize the linkages between collective memory, place attachment, and national identity (2003, 75). Indeed, the resources, time, and passion invested in the creation of expansive networks of national parks and monuments radiating from prominent government and cultural buildings concentrated in capital cities help condition the spatiality of national memory and homeland across multiple scales (Herb and Kaplan 2018).

Natural landscapes contribute to the contours of national identity, but the built environment has provided a preeminent influence shaping national mythscapes of memory, place attachment, and identity (Bilsel 2017; Johnson 2018; Post 2018; Diener and Hagen 2019). Authoritarian regimes in particular seem to place a special emphasis on maintaining the calculated control of the built environment and public space (Rifkin 2012; Bodenschatz 2014; Lewis 2015; Sevilla-Buitrago 2015). Historic architecture and commemorative monuments serve as prominent focal places of memory in these mythscapes, and the processes of historic preservation are deployed to deliberatively manipulate those places to advance specific ideological agendas. These geographies of historic preservation, whether at the scale of individual structures, neighborhoods, cityscapes, or regional/national landscapes, foster bonds between people and place through mutually reinforcing senses of collective memory and place attachment. Those bonds can be established and perpetuated through the immediacy of direct experiences of place, but the production and consumption of cultural artifacts —such as assorted visual mediums, literature, reenactments, or souvenirs—can also propagate senses of place and give credence to notions of collective memory, place attachment, and national identity. Cumulatively, these mythscapes help tie together what anthropologist Sharon Macdonald called a "memory-heritage-identity complex" (2013, 5).

Places of Nazi Memory and Identity

Building on scholarship on historic preservation, collective memory, and German national identity (see Confino 1997; Koshar 1998; Lekan 2004), this article examines the range of places of memory created, altered, and imagined by or on behalf of the Nazi movement. In doing so, this article explicates the types of places incorporated into Nazi commemorative initiatives, the regime's propensity to glorify generic figures instead of specific individuals, and its ideological motivations for those proclivities. Grounded in the pseudoscience of social Darwinism, this commemorative work contextualized Nazism within a sweeping history of racial struggle and sacrifice in service to the German nation. Nazi propagandists promoted that narrative relentlessly as ideological scaffolding for understanding the Party's past and present, and ultimately belief in its future. Nazi officials unleashed an unremitting barrage of speeches, ceremonies, exhibitions, publications, and newsreels, among other media, to promulgate a "Nazi pantheon of heroes" and "create an instant heritage industry of their own" (Lambert 2007, 525; Baird 1990; Behrenbeck 1996). Scholars have devoted considerable attention to the historiography of those commemorative endeavors, but much less attention has been paid to the regime's efforts to create places of memory supporting that ideological narrative. In fact, the regime constructed an increasingly insidious and pervasive archipelago of places of memory that grounded the story of National Socialism with materiality and spatiality. This burgeoning spatial framework of Nazi memory worked to situate Hitler's movement as the latest manifestation of a primordial racial ontology stretching back into the mists of Germanic prehistory.

The idea of continuous racial struggle and war was ubiquitous within this framework and worked to reinterpret German history as Nazi history and vice versa. Given the emphasis on racial struggle and war, the Nazi regime predictably extolled the virtues of obedience, heroism, and sacrifice and accordingly subsumed all German war dead as proto-Nazi martyrs. By extension, the regime quickly created or appropriated places associated with that narrative into a veritable cult of death and martyrdom. An introductory Hitler Youth textbook, for example, purposely blurred the distinction between fallen soldiers of World War I and fallen members of the Nazi movement:

> In reverence and silent readiness, we remember our dead—the dead of the war and the dead of our movement. … The names of the dead of our movement reverberate across the land, the names of the battlefield cemeteries that form the sacred ring around the Reich resonate in remembrance. In us, there is no grief, only obedience and readiness reign.

The passage conflates remembrance of the nation's fallen and the Party's dead while simultaneously conditioning readers to be loyal and obedient without hesitation. The textbook solidified this connection by asking readers to imagine these places as "sites of dying, sites of belief, holy places of the Reich" (Dörner 1939, 156). The prevalence of sacrifice and death could engender pessimism or fatalism, but the Nazi regime spun that as part of a broader trajectory of national redemption. In that way, these places of Nazi memory allowed the regime to lay claim to the past, present, and future.

In practice, this framework afforded considerable room for diffuse initiatives championed by a variety of actors so long as they were "working towards the Führer" (Kershaw 1993). This decentralized context fostered malleable combinations of consensus and contestation that incorporated an expanding constellation of places of memory into the regime's commemorative work. This was hardly surprising given that one popular educational primer for youth characterized Germany as "strewn with sites of memory"; each constituted "a piece of memory and tradition, a holy place of German heroism, German culture and spiritual splendor … places of pilgrimage of the nation where Germans experience the wonder of his nation and summon belief in the future" (von Schumacher 1935, 95–96).

Despite that variability, this article argues that the Nazi regime focused on three overlapping commemorative endeavors: creating new places of memory associated with the history of the Nazi movement, appropriating established places of memory associated with German history, and planning new places of memory that would henceforth fuse the history of the German nation and Nazi movement. In the process, the Nazi movement sought to blanket Germany with a thick web of "validating places," defined by geographer Jeffrey S. Smith as "places that serve as memorials to some past event, tragic or otherwise … locations where personal and group identity is reinforced" (2018, 9). These efforts aimed to produce a topography of collective memory that forged a sense of place attachment evocatively epitomized by the phrase "blood and soil," a popular Nazi slogan binding the German race to an ancestral and primordial homeland. Given that ideological imperative, the regime's commemorative topography tended to focus on places that could be associated with more anonymous, collective, and generic heroes—like the farmer, soldier, or mother—as idealized members of a timeless national community bound together through race and place—blood and soil—instead of focusing on a cannon of "great" individuals. The following sections highlight some places of memory promoted by the Nazi regime across those commemorative endeavors, based on extensive reviews of contemporary publications and archival sources (Figure 1).

Places Sacred to the National Socialist Movement

The Nazi movement made commemorating its history a priority as it gained popularity coinciding with the 1930 parliamentary elections and especially so after seizing power in 1933. This imperative stemmed from the Party's relatively recent beginnings, its improbable rise to power, and the murky origins of its Führer. As a result, Nazi officials scrambled to create a commemorative topography across Germany through memorial plaques, historical monuments, and other types of markers. It was equally important that Germans experienced these nascent places of memory through rallies, rituals, and other performances, in effect creating a Nazi commemorative liturgy to foster bonds between people, places, and Party.

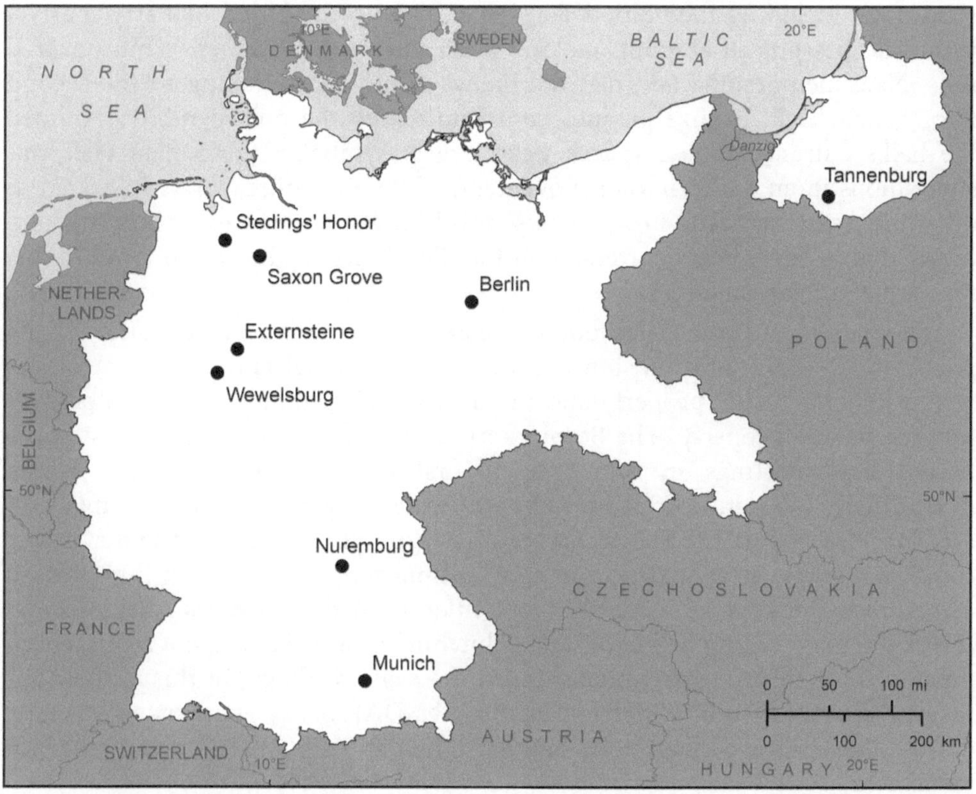

FIG. 1—Reference map of Germany in its 1937 borders showing the main locations discussed in the text. Source: Map created by James Leonard.

Naturally, places associated with Hitler would become central locations in this new national mythscape. Munich was the heart of this burgeoning cult of Hitler worship. Hitler's first residence in Munich was a nondescript apartment above a tailor's shop. Hitler left for military service the following year, but eventually returned to Munich and rented another unassuming apartment. After seizing power, Nazi officials mounted modest commemorative plaques on both buildings to mark their significance. A more monumental plaque was installed inside the Hofbräuhaus beer hall to remind visitors: "From this place, Adolf Hitler proclaimed on 24 February 1920 the program of the National Socialist German Workers' Party."

The movement's first substantive commemorative building project followed the purchase of a bourgeois residence in an upscale Munich neighborhood in 1930 (Bauer and others 1995; Heusler 2008). Hitler had the building remodeled as the official Party headquarters, dubbed the Brown House, but much of the interior served a commemorative function. The vestibule was lined with early

Nazi flags, including the "Blood Flag" stained with the blood of Party martyrs from the 1923 putsch attempt, making it arguably the movement's most sacred relic. Nazi memorabilia adorned the hallways and stairs leading to the second floor Senate Hall. Bronze plaques commemorating the putsch martyrs framed the hall's entrance. These spaces were largely symbolic; the Senate Hall was superfluous in an authoritarian movement, while most bureaucrats relocated to other offices by 1937. The Brown House was basically a showcase of Nazi history, rather than a Party headquarters, with the Blood Flag as the crown jewel in this repository of Nazi memory.

The regime had greater latitude to create commemorative places after seizing power in 1933 and by extension control over municipal and provincial offices responsible for public property and urban/regional planning, as well as permitting for private property. The Sterneckerbräu brewery and beer hall hosted the Party's first meetings in 1919 and its first office in a small back room. Accordingly, the nondescript brewery was widely regarded as the birthplace of the Nazi movement. The Sterneckerbräu was promptly redecorated so that Hitler could inaugurate it as a new museum in time for the November 1933 putsch commemorations. A large portrait of Hitler framed by swastika flags peered across the main dining hall at an eagle clutching a swastika insignia mounted on the opposite wall. An inscription flanked the eagle reading: "In this corner, our Reich Chancellor Adolf Hitler founded the NSDAP on 24 February 1920. Here, the first seven brave champions laid the foundation for the German freedom movement."

The backroom was carefully restaged as the first Party office, allegedly featuring original furniture and equipment like the Party's typewriter and the cigar box that safeguarded membership dues. Party leaflets, posters, and other accessories suggested the office was frozen in time since those early days. This space would figuratively transport visitors back in time to experience the movement's humble roots and the headquarters for its "first major battles." The museum's guidebook stressed the experiential immediacy of place: "one has to be there oneself in order to feel a hint from the time out of which the great movement was created" (Schüßler 1935, 5, 11).

The Bürgerbräukeller was a more popular venue for later Party meetings. This beer hall gained added notoriety as the launching point for Hitler's unsuccessful putsch in November 1923. After storming the hall and declaring revolution, Hitler and his supporters eventually marched through downtown Munich before heading toward the Bavarian defense ministry. Police officers blocked the putschists near the Field Generals' Hall. The ensuing gunfire killed four officers and sixteen marchers. Hitler's revolt collapsed quickly, but nonetheless provided an initial pantheon of Party martyrs.

The Field Generals' Hall rapidly became a central location in Nazi lore, but the Party had no means to mark the location until 1933 when Hitler ordered a bronze slab topped by an eagle clutching a swastika erected on the side of the hall. One side praised the martyrs for dying in "true belief in the resurrection of their nation," while the other side exclaimed "and yet you have triumphed!" The idea of triumph over or through death has obvious parallels to Christianity's belief in the resurrection of Jesus. A SS honor guard added to the solemnity, as did the requirement that passersby give the Nazi salute.

The Field Generals' Hall became one of the movement's most ubiquitous icons, but Hitler found it underwhelming. As a result, Hitler ordered the redesign of the famous Königsplatz square, located a short distance from the Brown House. Hitler already planned to build additional Party offices in the area, but the project expanded significantly during 1933. The three main structures surrounding the Königsplatz were not altered, but Hitler had the fourth side lined with two pairs of matching buildings. The largest structures were a pair of identical office buildings, but the redesigned space focused on two Temples of Honor sitting astride the main axis. The temples were square, open-air atria about seven meters tall, intended as the final resting places for the putsch martyrs. Hitler ordered the plaza covered with granite pavers to better accommodate rallies. These changes, especially the presence of the martyrs, radically reconfigured this public space into a "forum of the movement" expressly for Party commemorations. Indeed, the Party press extolled the redesigned Königsplatz as "a precinct of dignity and majesty consecrated through the remembrance of sacrifice, an Acropolis Germaniae" (Heilmeyer 1935, 140–141). These overt allusions to classical architecture reflected Hitler's personal fascination with the grandeur ancient Greece and Rome and desire to achieve comparable glory for his thousand-year Reich (Scobie 1990; Hagen 2010).

As the scene of the 1923 putsch, home of the Brown House Party headquarters, and officially decreed as the "capital of the movement" in 1935, Munich hosted the most extensive commemorative program, and unlike elsewhere, most of projects reached completion within a relatively short time. This allowed Party officials to implement a robust liturgy linking these places of memory. The putsch became a central component in the Party's commemorative calendar and served to advance the themes of struggle and sacrifice on behalf of the movement (Hockerts 1993; Lehmbruch 1995; Pappert 2001; Köpf 2005). The inaugural ceremony occurred in November 1933. Hitler opened the occasion with an evening speech at the Bürgerbräukeller. The following day, Hitler and his companions retraced the 1923 march route, which was decorated with a profusion of flags, banners, and pylons topped with flaming cauldrons (Figure 2). Each pylon bore a martyr's name, which was announced by loudspeaker as Hitler passed. The procession eventually reached the Field Generals' Hall, decorated with 16 commemorative pylons, where Hitler laid a wreath to conclude the ceremony.

9. November: Erinnerung und neues Gelöbnis

FIG. 2—This highly stylized propaganda photo depicted Hitler and his entourage re-enacting the November 1923 putsch march. The caption reads: 9 November: Memory and A New Vow.
Source: Fiehler, K., ed. 1937. *München baut auf: Ein Tatsachen- und Bildbericht über den nationalsozialistischen Aufbau in der Hauptstadt der Bewegung.* Munich: Franz Eher, np.

The reenactment expanded in 1935 to incorporate the recently completed Temples of Honor. Sixteen ornate sarcophagi containing the martyrs spent the night under an honor guard at the Field Generals' Hall. Hitler and his entourage retraced the route from the Bürgerbräukeller to the Field Generals' Hall, but in a new twist, the procession then transported the sarcophagi to the Königsplatz for interment in the new temples. Thousands of Nazi troopers standing in rigid formations filled the square (Figure 3). The troopers hollered "Here!" as the names of the martyrs were called out during an honorary roll call before the sarcophagi were placed in the temples. The expanded ceremony conveyed a clear message. Retracing the original route to the Field Generals' Hall commemorated the martyrs and other hardships endured by the movement during its struggle for power. The continued march to the temples represented the movement's triumph and the symbolic redemption of the martyrs. To conclude the commemorations, SS recruits later swore allegiance to Hitler in front of the Field Generals' Hall, effectively promising to martyr themselves.

FIG. 3—Hitler and his entourage arrive at the Königsplatz, already packed with formations of Nazi troopers, during the 1935 putsch commemorations. The two Temples of Honor are in the center top with the Brown House partially visible directly behind the temple on the left.
Source: Fiehler, K., ed. 1937. *München baut auf: Ein Tatsachen- und Bildbericht über den nationalsozialistischen Aufbau in der Hauptstadt der Bewegung*. Munich: Franz Eher, 65.

Marking a high point among public holidays declared by the Nazi regime, this carefully calculated arrangement of places, processions, and accoutrements subsumed marchers and observers into a spatial topography of Nazi memory extolling the virtues of heroism and sacrifice as epitomized by the putsch martyrs. Cumulatively, these places, from humble apartments to grandiose Temples of Honor, traced Hitler's ascent to become Germany's undisputed leader and attracted a growing number of visitors. As Mayor Karl Fiehler claimed: "The mighty buildings of the Führer, the new reputation of Munich as the Capital of the Movement with its historical places from the *Kampfzeit* [time of struggle], and the announcement of the city as the City of German Art have attracted numerous domestic and foreign visitors" (1937, 49). Guidebooks soon appeared to direct the Party faithful and the curious to this mythscape of "places sacred to the National Socialist Movement," as designated in a guidebook commissioned by Fiehler (Spahn 1937, np). Countless other items, including photographs, newsreels, radio broadcasts, and assortment memorabilia promulgated the regime's mythology of 1923 and the broader cult of sacrifice and martyrdom.

The tireless work of Nazi propagandists inspired officials across Germany to imitate the Munich commemorations on smaller scales. The veneration of the 1923 martyrs remained the central theme, commonly enshrined in the entryway of local Party offices, but localities also featured their particular contributions to the movement's rise to power. In Berlin, for example, two SA members published a guidebook to assorted Nazi hangouts, locations of street fighting, and places of Nazi martyrdom, and in doing so recast that metropolis as "National Socialist Berlin" (von Engelbrechten and Volz 1937). The same impulses gave rise to a burgeoning campaign to rename streets, squares, and other public places in Hitler's honor. Germany soon had numerous Adolf Hitler streets, as well as places named for other Nazis, both living (Hermann Göring) and deceased (Horst Wessel). Hitler worried this rapid proliferation trivialized the Party and so ordered that honorary names be limited to new places with few exceptions. Ordinances in 1933 codified criteria for renaming places and other commemorative markers, instructing officials to "refrain from altering historical markers" because "only that which the national revolution has itself built for the future can carry its name and the names of its leading figures" (Meier 1933, 231). Consequently, officials looked to new construction for commemorative opportunities, such as streets named after the putsch martyrs in Munich's new Ramersdorf suburb, to promulgate this new national mythscape binding people, Party, and place.

A Great Castle of the Dead

The Nazi regime simultaneously endeavored to rewrite all of Germany history as a story of racial struggle anchored in space and place. This reinvention of history obviously involved rewriting history textbooks. The regime's efforts to rewrite Germany's spatial history, that is to redesign the country's spaces and places of collective memory, were equally important and concomitantly appropriated established place attachments and identities. Building this new national mythscape required a substantial amount of work in some cases. In other cases, relatively minor alterations accomplished the basic objectives. In both instances, the regime allowed for a considerable amount of local initiative, at least so long as the effort aligned with Party ideology.

Aside from the 1923 putsch, World War I was the single most pivotal event in Nazi commemorative lore. Many Party members served in the war, and the cataclysm of defeat provided the catalyst for the Party's founding. Nazi propaganda cast defeat as symptomatic of decades of cultural degeneration, racial miscegenation, and weak leadership while simultaneously promising Germany's eventual rejuvenation and redemption through Hitler's movement. It was predictable then that the war figured prominently in Party propaganda and commemorations. The incorporation of the Hall of Honor into the Nuremberg rallies provided an early example. Built to honor soldiers killed in World War I, the hall remained unchanged, but Hitler ordered the surrounding parkland redesigned as an expansive arena that could accommodate massive formations of Nazi troopers enclosed by grandstands for thousands of spectators. Bookended by two massive eagle statues, the main tribune centered on Hitler's speaking podium, accented by a backdrop of three towering swastika banners. To culminate the ceremony, Hitler strode from the tribune along a broad granite path cut through the Party formations to reach the hall on the opposite side of the arena. There, Hitler honored the war dead and the Party's martyrs, effectively claiming the world war dead as part of the Nazi movement.

The regime also took substantive action regarding the Tannenberg monument in East Prussia. Veteran groups raised funds to build a monument commemorating Germany's victory in 1914 over Russian forces. The victory gained special significance as the only major World War I battle on German territory, among other reasons, and therefore afforded the regime direct control over the battlefield. The octagonal monument featured stout towers standing approximately 20 meters high at the center of each side, giving the structure a castlelike appearance. The interior courtyard featured four paths that converged on a central platform topped by a bronze cross over the remains of 20 unknown soldiers.

The monument was dedicated in 1927, but Hitler showed little interest until the death of President Paul von Hindenburg, the famed commander at Tannenberg, in 1934. Hitler ordered the monument renovated as Hindenburg's mausoleum. The main structure and towers remained largely unchanged, except for the addition of Hindenburg's new crypt below the tower opposite the main entrance, but the courtyard was excavated to lower the ground level by over two meters with staircases encircling this new depression. The entire courtyard was then paved with an iron cross of darker stone marking the center (Figure 4).

Overall, the redesigned Tannenberg resembled an octagonal amphitheater of stone surrounded by ramparts and towers.

The renovations were completed by October 1935, allowing Hindenburg, his wife, and the unknown soldiers to be entombed in their intended final resting places. The dedication ceremony featured extensive pomp and publicity as Hitler expropriated this rechristened Reich Honorary Monument Tannenberg. In tandem with the interment of the putsch martyrs in Munich, Hitler and the Party had largely subsumed Germany's World War I dead within the memory of the movement, a process that would rapidly extend as the rest of German history was recast as a story of perpetual racial struggle and sacrifice culminating in the Nazi movement. One of Hitler's confidants tellingly described the redesigned Tannenberg as a "great castle of the dead" (Troost 1938, 34).

The experience of other monuments was varied. Hitler ordered the Victory Column, a monument to German unification in 1871, relocated from in front of the Reichstag parliament to the Tiergarten park as part of his plans to redesign Berlin. In the process, Hitler had the column's height increased by around 7.5 meters to be even more imposing (Figure 5). In other cases, Nazi officials coopted existing monuments without any changes other than perhaps adding swastika flags. For example, Nazi officials incorporated the gargantuan

FIG. 4—Hitler had the interior courtyard of the Tannenberg monument lowered and paved so the space would be better suited for Party rallies.
Source: Rittich, W. 1938. *Architektur und Bauplastik der Gegenwart.* Berlin: Rembrandt, 67.

FIG. 5—Hitler ordered the Victory Column relocated to the Tiergarten park as part of his plans to redesign Berlin. Hitler also ordered the addition of the fourth top section of the main column to make the monument taller. This photograph shows the relocated and altered monument at its current location. Source: Photograph by Joshua Hagen.

Monument to the Battle of the Nations, which commemorated victory over French forces near Leipzig in 1813, into Party assemblies without structural changes. This variability was indicative of a certain degree of hesitancy and ambiguity evident in the regime's engagement with established places of memory and place attachments that lacked clear direct or even indirect ties to the Nazi movement and its ideology.

THE MEANING OF STRUGGLE, SACRIFICE, AND VICTORY

The Nazi regime also sought to create new places of memory, but unlike the places associated with Hitler or seminal Party events, these sites had no prior connection to the movement or German history. For example, numerous communities established place names to honor Wessel or other Party martyrs, even though they lacked connections to those localities. Other communities erected more general monuments to the Party, mostly as local initiatives. These monuments tended to feature traditional commemorative elements like columns, plinths, and statues bearing commemorative inscriptions. These monuments materialized quickly. In April 1933, officials in the small town of Gunzenhausen, about 40 kilometers southwest of Nuremberg, unveiled one of the first Party monuments, a simple, seven-meter-tall obelisk. Not far away in Rothenburg ob der Tauber in September 1934, local officials dedicated a square column inscribed with a swastika and the standard Nazi trope "Germany has awakened!" The six-meter-tall column, topped by a bronze eagle with outstretched wings, was unveiled amid great pageantry. Countless localities across Germany undertook similar efforts to memorialize the Party in public space and forge new bonds between people, place, and Party.

Officials also promoted the construction of open-air amphitheaters as new places of memory. By the mid-1930s, as many as 400 amphitheaters were planned across Germany, commonly in areas with dramatic or picturesque scenery. In 1937, Nazi propaganda minister Joseph Goebbels described these places as the "political churches of National Socialism" (Stephan 1937, 1193–1197), but the campaign was winding down with less than one-fifth of planned projects completed. Most of the completed amphitheaters languished in relative obscurity, largely due to their remote locations and poor productions, but a few achieved broad popularity.

The Stedings' Honor amphitheater near Bremen was perhaps the most popular. The archbishop of Bremen ordered a crusade in 1234 against the peasants of the Stedingen district for allegedly refusing to pay taxes and other feudal obligations. The Stedingers resisted vigorously but were ultimately massacred. The episode was little known beyond the immediate vicinity until Party boss Carl Röver ordered the construction of a new amphitheater to commemorate the 700 anniversary of the conflict. The amphitheater featured semicircular rows of seating banked into the ground surrounding a central stage, which featured a faux peasant village replete with homey cottages, rustic farm buildings, and a quaint stone church, as well as a moat separating the stage from the seating area.

The project garnered support from prominent Nazis, most notably Heinrich Himmler, Alfred Rosenberg, and Richard Walther Darré. The play debuted to sold out audiences during 1935 and proved popular in subsequent years. The play recounted the story from the perspective of the Stedingers, who were depicted as honorable peasants defending their community against overwhelming odds. Röver hoped to capitalize on the play's popularity by building out the theater into a new "Cult Site Stedingen," although the outbreak of war prevented additional construction (Finsterhölzl 1999). Nonetheless, the play and its staging constituted a new place of memory glorifying the idea of heroism and sacrifice on behalf of the national community.

Those same impulses fueled the creation of the Saxon Grove memorial near Verden. Popular legend held that Charlemagne executed 4,500 Saxon warriors in 782 as part of his campaign to subjugate northeastern Germany and convert the populace to Christianity. Himmler and Rosenberg backed a project to create a new place of memory for this Massacre of Verden. The site would feature 4,500 large stones lining pathways leading to a small clearing in a clump of trees alleged to be the spot of the executions. Five rustic farmhouses were relocated to the grove to create a small hamlet as a SS training facility. Himmler, Rosenberg, and Darré attended the grove's dedication during the 1935 summer solstice (Kaldewei 2006). Given that the Nazi Party positioned its history as a story of struggle, sacrifice, and martyrdom on behalf of a raced-based national community, it is hardly surprising that those same themes permeated its commemorative work regarding German history stretching back to prehistoric times. It is also noteworthy that many of these sites—Tannenberg, Stedings' Honor, Saxon Grove—immortalized and sanctified resistance and death in the face of foreign invaders.

Himmler, Rosenberg, and Darré were also keenly interested in Germanic prehistory. Himmler, for example, commissioned numerous excavations focused on pagan Germanic peoples and took a particular interest in the Externsteine, a row of unusual vertical rocky outcroppings in Westphalia. Himmler ordered extensive excavations to prove the Externsteine was an important place of pagan Germanic worship, perhaps even the fabled *Irminsul* shrine supposedly destroyed by Charlemagne. The excavations failed to uncover definitive ties to pagan Germans. Undeterred, Himmler developed plans to turn the area into a sacred memorial grove centered on the Externsteine and thereby incorporate the place's supposed pagan mysticism into the new religion of National Socialism, although little was realized.

The Externsteine's location in northwestern Germany carried extra significance for Himmler and like-minded Nazi ideologues who regarded this general region as the protohomeland of various Germanic tribes that eventually expanded to the south and east to form the German people. Given that, Himmler was eager to establish a presence across the region and so took possession of the nearby Wewelsburg Castle in 1934 with the intent of transforming it into the nucleus of a sprawling SS complex. Himmler's architects took free rein to rebuild and remodel the castle—augmented by multiple nearby auxiliary structures—into the centerpiece of a training academy, research institute, and leadership retreat grounded in SS ideology and pseudoscience.

Cumulatively, Himmler and other like-minded Nazis attempted to lay claim to Germany's prehistoric past to substantiate central tenets of Party ideology through the calculated creation of an archipelago of places of memory across northwestern Germany supporting their racialized interpretation of history. These examples illustrate the Nazi movement's sustained efforts to infuse German history with that racially charged ideology. These new places of memory and place attachments were to become sites of pilgrimage and reverence for the German people, political churches of National Socialism, while advancing belief in the necessity of racial struggle and war for the survival of the national community.

Hitler allegedly scorned his lieutenants' passion for delving into Germanic prehistory, and Himmler's zeal for prehistoric archaeology in particular, according to top Nazi Albert Speer (1981, 994–995). Hitler was not dismissive of all memory work though; he simply envisioned it on a grander and more modern scale, even compared to earlier commemorative efforts in Nuremberg and Munich. Berlin would be the focal point of future Nazi commemorations as part of Hitler's dreams to thoroughly redesign the capital around spacious plazas and axial boulevards punctuated by gargantuan monuments featuring a ponderous and pretentious architectural style developed by Hitler and Speer. The centerpiece of Hitler's plan was a grand north-south boulevard—stretching approximately seven kilometers—connecting the main government district with a new train station. Travelers exiting the station would immediately encounter an expansive rectangular plaza flanked by war trophies stretching to the north. A colossal triumphal arch, dwarfing the Arc de Triomphe in Paris, would dominate the north end of the plaza. Reaching around 120 meters high, Hitler's arch would have ample surfaces to bear the names of the 1.8 million German soldiers fallen during World War I.

The grand boulevard would pass the Soldier's Hall near its northern terminus. Designed by Wilhelm Kreis, one of Germany's most respected elder architects, the hall's interior was composed of two main spaces, both overwhelming in scale. The main level consisted of a bright, voluminous lobby beneath a towering barrel-vaulted ceiling, while the lower level Funeral Hall was a gloomy, groin-vaulted crypt (Figure 6). Praised as "a new, true German cathedral," the Soldier's Hall was less about commemorating those who had fallen in battle and more about conditioning the living to prepare for martyrdom on behalf of the Party. Thus, in the words of one of Speer's top architects, the hall spoke "more of the vows of the living as of the legacy of the dead" (Tamms 1943, 57).

These same themes figured prominently in Hitler's plans for a massive obelisk in Munich to commemorate the triumph of nationalist troops over communist militias in 1919. The planning progressed through successively larger designs until the last plan in 1944 called for an obelisk towering up to 250 meters high topped by an eagle with outstretched wings. By this time, the obelisk had also evolved into a more general Monument to the Movement. This evolution again highlighted the tendency to conflate all of German history and heroes, at least as understood within Nazi ideology, as somehow tied to the movement. Beyond their obvious commemorative functions, these massive monuments would also become centerpieces of broader urban landscapes that framed colossal displays of Party power and pageantry befitting a thousand-year Reich. Other than a hulking concrete test footing for the triumphal arch, none of these grandiose projects progressed beyond the planning stages.

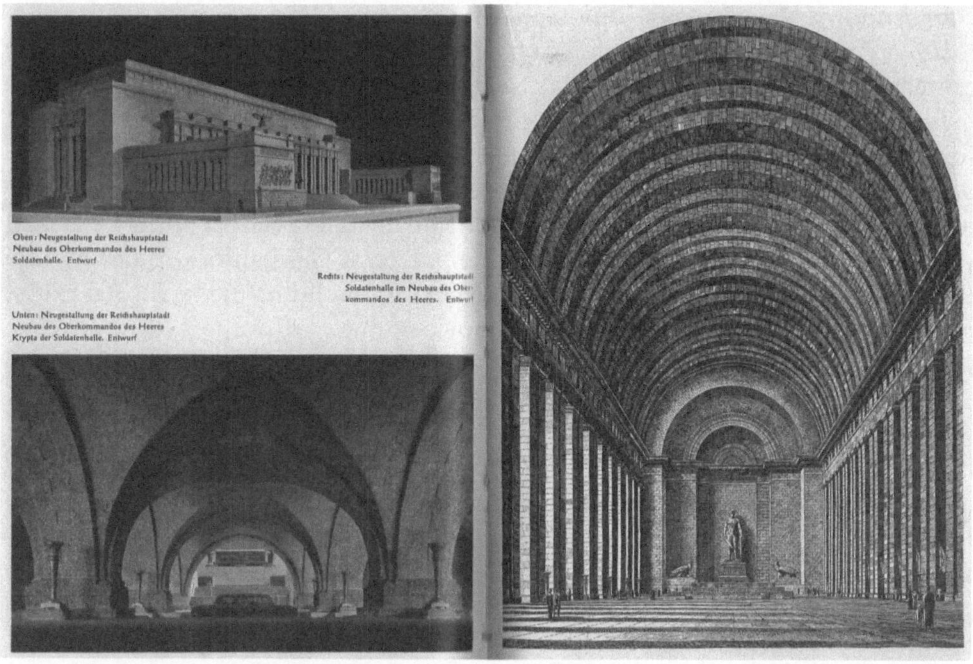

FIG. 6—Designed by Wilhelm Kreis, the Soldier's Hall featured - clockwise from top-left - a hulking neoclassicist exterior, an bright airy lobby, and gloomy cavernous crypt.
Source: Stephan, H. 1944. *Wilhelm Kreis*. Oldenburg: Gerhard Stalling, 72-73.

Hitler's decision to invade Poland in 1939 guaranteed that the regime would launch one final macabre commemorative initiative to ensure that remembrances of Germany's World War II dead unfolded in accordance with Nazi ideology. In 1941, Hitler commissioned Kreis to design a series of massive memorials and burial complexes for the current war. Kreis soon produced sketches for enormous cenotaphs to commemorate major World War II battlefields and thereby delineate the territorial extent of Hitler's thousand-year Reich "from Narvik to Africa, from the Atlantic to the plains of Russia, beginning at the borders of the Greater German Reich and ending on the front lines of the greatest battle in the world" (Tamms 1943, 51).

Accentuating their massive proportions, the cenotaphs would dominate the surrounding landscape and often give the impression of hilltop citadels. Traditional monuments, like arches, columns, and obelisks, tended to be structures that people gathered in front of or around. In contrast, visitors would generally enter these cenotaphs through successive spaces leading to a central sunken crypt. In that sense, Kreis' designs retained many of the basic features of Tannenberg and other world-war memorials that conveyed "the deep, mysterious, harrowing impression of a Romanesque crypt, thereby renewing the connection of the warriors' cenotaph with the neighboring graves" (Gstettner 1940, 154). Given its vast scale, Kreis logically planned to build the largest cenotaph somewhere on the Russian steppe. Shaped like a conical pyramid, the monument would stand about 150 meters tall to encompass an interior crypt soaring about 100 meters high that would awe and engulf visitors (Figure 7). One of Hitler's inner circle confidently declared of Kreis' memorials that "the meaning of struggle, sacrifice, and victory will find in them eternal form in lasting stone" (Troost 1943, 7). Surrounded by the graves of Germany's war dead, these cenotaphs would constitute new places of memory outlining the territorial extent of Hitler's new empire and its concomitant national mythscape.

Conclusion: Places for the Holy Dead

Kreis' sketches marked the culmination of the regime's efforts to cement pivotal historical events, especially the world wars, within a pantheon of Nazi heroes and sacrifice. By 1942, the distinction between the Party's martyrs and Germany's war dead had virtually disappeared. Official guidelines for commemorating the 1942 putsch anniversary explained that "the 9th November unites the entire national community in remembrance of the dead of the movement, but also in remembrance of the fallen of the First World War and the current war … The sixteen dead of the Field Generals' Hall and all blood victims of the movement are, therefore, joined in a firm connectedness with the fall heroes of this war" as part of the Nazi Party's pantheon of "holy dead" (Zum 9. November 1942, 492, 498).

The Nazi movement was new and the bulk of its members joined shortly before 1933. The Nazi leadership sought to construct a long and venerable pedigree that positioned the movement as the path to ultimate redemption in an enduring racial struggle for national survival. It was relatively easy to rewrite history books accordingly, but the regime sought other mechanisms to initiate new recruits into a pseudoreligious dogma grounding Nazi ideology back through German history. A new slate of commemorative places and practices emerged to serve that need beginning with places associated with Hitler and the Party's rise to power, but soon swelled to include an assortment of real and imagined historical places that could be associated with national mythscape of racial struggle, sacrifice, and martyrdom. As they proliferated across Germany and beyond, these places of memory were intended to convey a pervasive sense of national struggle with the Party at the vanguard, and of course, requiring unquestioning loyalty and belief. The Nazi Party exploited these places of memories to promulgate a new cult of death in which remembrances of the past would condition Germans as martyrs in waiting for the movement. In that framework, the notion of undifferentiated masses of followers dedicated to the movement and willing to lay down their lives led Nazi commemorative projects to focus on places of memory the glorified the sacrifices of "generic" heroes, like soldiers or farmers, with the obvious exception of Hitler as Führer.

FIG. 7—Wilhelm Kreis designed a series of massive cenotaphs to mark the borders of Hitler's empire. The largest cenotaph, seen here, was to be located on the Russia steppe somewhere near the Dnieper. Source: Stephan, H. 1944. *Wilhelm Kreis.* Oldenburg: Gerhard Stalling, np.

The Nazi movement provides an extreme example of the insidious potential of places of memory and place attachment, but simultaneously suggests the broad applicability of the person-process-place framework. Indeed, scholars have made significant strides in refining theoretical and methodological approaches to understanding the varied dynamics of place attachment over the last decade. Historic preservation offers a fruitful avenue to continue that progress. Historic preservation encompasses a set of processes that bind together people and place to produce shared memories and experiences of the past. The processes of historic preservation provide a means to produce places of memory that foster "particularly poignant encounters with place" (Strong 2018, 20) and thereby confer "culturally shared emotional/affective meanings to a particular space or piece of land" (Low 1992, 165). As a result, these places of memory engender notions of place attachments and simultaneously serve as "validating places" (Smith 2018, 9) that condition specific mentalities and behaviors in support of group identity at multiple scales and registers. In that sense, historic preservation elucidates the nexus of people, place, and process that contributes to places of memory and place attachment in both historical and contemporary contexts.

References

Alderman, D. H., and J. F. J. Inwood. 2013. Landscapes of Memory and Socially Just Futures. In *The Wiley-Blackwell Companion to Cultural Geography*, edited by N. C. Johnson, R. H. Schein, and J. Winders, 186–197. Malden, Mass.: John Wiley.

Assmann, A. 2009. *Erinnerungsräume: Formen und Wandlungen des kulturellen Gedächtnisses.* Munich: C. H. Beck.

Assmann, J. 2011. Communicative and Cultural Memory. In *Cultural Memories: The Geographical Point of View*, edited by P. Meusburger, M. Heffernan, and E. Wunder, 15–27. New York: Springer.

Baird, J. 1990. *To Die for Germany: Heroes in the Nazi Pantheon.* Bloomington: Indiana University Press.

Bauer, R., and others, eds. 1995. *München: "Hauptstadt der Bewegung": Bayerns Metropole und der Nationalsozialismus.* Munich: Klinkhardt und Biermann.

Behrenbeck, S. 1996. *Der Kult um die Toten Helden: Nationalsozialistische Mythen, Riten und Symbole 1923 bis 1945.* Vierow: SH-Verlag.

Bell, D. S. A. 2003. Mythscapes: Memory, Mythology, and National Identity. *British Journal of Sociology* 54 (1): 63–81.

Bilsel, C. 2017. Architecture and the Social Frameworks of Memory: A Postscript to Maurice Halbwachs' 'Collective Memory.' *International Journal of Architecture & Planning* 5 (1): 1–9.

Bodenschatz, H. 2014. Urban Design for Mussolini, Stalin, Salazar, Hitler and Franco. *Planning Perspectives* 29 (3): 381–92.

Confino, A. 1997. *The Nation as a Local Metaphor: Württemberg, Imperial Germany, and National Memory, 1871–1918.* Chapel Hill: University of North Carolina Press.

Connerton, P. 1989. *How Societies Remember.* Cambridge, U.K.: Cambridge University Press.

De Nardi, S., H. Orange, S. High, and E. Koskinen-Koivisto, eds. 2020. *The Routledge Handbook of Memory and Place.* London: Routledge.

Diener, A. C., and J. Hagen, eds. 2019. *The City as Power: Urban Space, Place, and National Identity.* Lanham, Md.: Rowman & Littlefield.

Dörner, C. 1939. *Das Deutsche Jahr: Feiern der Jungen Nation.* Munich: Franz Eher.

von Engelbrechten, J. K., and H. Volz. 1937. *Wir wandern durch das nationalsozialistische Berlin.* Munich: Franz Eher.

Erll, A. 2005. *Kollektives Gedächtnis und Erinnerungskulturen: Eine Einführung.* Stuttgart: J. B. Metzler.

Fiehler, K., ed. 1937. *München baut auf: Ein Tatsachen- und Bildbericht über den Nationalsozialistischen Aufbau in der Hauptstadt der Bewegung.* Munich: Franz Eher.

Finsterhölzl, C. 1999. Die Einweihung der 'Niederdeutschen Gedenkstätten' 'Stedingsehre': Ein Beispiel der nationalsozialistischer Selbstinszenierung im Gau Weser-Ems. *Oldenburger Jahrbuch* 99: 177–205.

Foote, K. E., and M. Azaryahu. 2007. Toward a Geography of Memory: Geographical Dimensions of Public Memory and Commemoration. *Journal of Political & Military Sociology* 35 (1): 125–144.

Giaccaria, P., and C. Minca, eds. 2016. *Hitler's Geographies: The Spatialities of the Third Reich.* Chicago: University of Chicago Press.

Gstettner, H. 1940. Die deutsche Gestaltung des Kriegergrabes. *Kunst für Alle* 55 (77): 144–154.

Hagen, J. 2010. Architecture, Symbolism, and Function: The Nazi Party's 'Forum of the Movement.' *Environment and Planning D: Society and Space* 28 (3): 397–424.

Hagen, J., and R. C. Ostergren. 2020. *Building Nazi Germany: Place, Space, Architecture, and Ideology.* Lanham, Md.: Rowman & Littlefield.

Halbwachs, M. 1992. *On Collective Memory.* Translated by L. A. Coser. Chicago: University of Chicago Press.

Heilmeyer, A. 1935. Die Stadt Adolf Hitlers. *Süddeutsche Monatshefte* 33: 140–141.

Herb, G. H., and D. H. Kaplan, eds. 2018. *Scaling Identities: Nationalism and Territoriality.* Lanham, Md.: Rowman & Littlefield.

Heusler, A. 2008. *Das braune Haus: Wie München zur 'Hauptstadt der Bewegung' wurde.* Munich: Deutsche Verlags-Anstalt.

Hockerts, H. G. 1993. Mythos, Kult und Feste: München im nationalsozialistischen 'Feierjahr'. In *München: "Hauptstadt der Bewegung": Bayerns Metropole und der Nationalsozialismus*, edited by R. Bauer, and others, 331–341. Munich: Klinkhardt und Biermann.

Hoelscher, S., and D. H. Alderman. 2004. Memory and Place: Geographies of a Critical Relationship. *Social & Cultural Geography* 5 (3): 347–355.

Jaskot, P. B. 2000. *The Architecture of Oppression: The SS, Forced Labor and the Nazi Monumental Building Economy.* New York: Routledge.

Johnson, C. 2018. The Changing Context of German Nationhood. In *Scaling Identities: Nationalism and Territoriality*, edited by G. H. Herb and D. H. Kaplan, 51–65. Lanham, Md.: Rowman & Littlefield.

Jones, O., and J. Garde-Hansen. 2012. *Geography and Memory: Explorations in Identity, Place, and Becoming.* Basingstoke: Palgrave Macmillan.

Kaldewei, G. 2006. *'Stedlingsehre' soll für ganz Deutschland ein Wallfahrtsort werden: Dokumentation und Geschichte einer NS-Kultstätte auf dem Bookholzberg 1934–2005.* Delmenhorst: Aschenbeck & Holstein.

Kershaw, I. 1993. 'Working Towards the Führer': Reflection on the Nature of the Hitler Dictatorship. *Contemporary European History* 2 (2): 103–118.

Köpf, P. 2005. *Der Königsplatz in München: Ein deutscher Ort.* Berlin: Ch. Links.

Koshar, R. 1998. *Germany's Transient Pasts: Preservation and National Memory in the Twentieth Century.* Chapel Hill: University of North Carolina Press.

Lambert, P. 2007. Heroisation and Demonisation in the Third Reich: The Consensus-building Value of a Nazi Pantheon of Heroes. *Totalitarian Movements and Political Religions* 8 (3–4): 523–546.

Lehmbruch, H. 1995. Acropolis Germaniae: Der Königsplatz—Forum der NSDAP. In *Bürokratie und Kult: Das Parteizentrum der NSDAP am Königsplatz in München*, edited by I. Lauterbach, J. Rosenfeldt, and P. Steinle, 17–45. Munich: Deutscher Kunstverlag.

Lekan, T. M. 2004. *Imagining the Nation in Nature: Landscape Preservation and German Identity, 1885–1945.* Cambridge, Mass.: Harvard University Press.

Lewicka, M. 2011. Place Attachment: How Far Have We Come in the Last 40 Years? *Journal of Environmental Psychology* 31 (3): 207–230.

_____ 2014. In Search of Roots: Memory as Enabler of Place Attachment. In *Place Attachment: Advances in Theory, Methods and Applications*, edited by L. C. Manzo and P. Devine-Wright, 49–60. London: Routledge.

Lewis, D. 2015. "Illiberal Spaces:" Uzbekistan's Extraterritorial Security Practices and the Spatial Politics of Contemporary Authoritarianism. *Nationalities Papers* 43 (1): 140–159.

Low, S. M. 1992. Symbolic Ties that Bind: Place Attachment in the Plaza. In *Place Attachment*, edited by I. Altman and S. M. Low, 165–185. New York: Plenum.

Macdonald, S. 2013. *Memorylands: Heritage and Identity in Europe Today.* London: Routledge.

Maus, G. 2015. Landscapes of Memory: A Practice Theory Approach to Geographies of Memory. *Geographica Helvetica* 70 (3): 215–223.

Meier, B. 1933. Zur Umbenennung historischer Strassennamen. *Deutsche Kunst und Denkmalpflege* 35 (6): 231.
Meusburger, P., M. Heffernan, and E. Wunder, eds. 2011. *Cultural Memories: The Geographical Point of View*. New York: Springer.
Nora, P. 1989. Between Memory and History: Les Lieux de Mémoire. *Representations* (26): 7–24.
Pappert, L. 2001. *Der Hitlerputsch und seine Mythologisierung im Dritten Reich*. Neuried: ars una.
Post, C. 2018. Making Place through the Memorial Landscape. In *Explorations in Place Attachment*, edited by J. S. Smith, 83–96. London: Routledge.
Rifkin, D. 2012. "Everything in the State, Nothing Against the State, Nothing Outside the State:" Corporativist Urbanism and Rationalist Architecture in Fascist Italy. *Planning Perspectives* 27 (1): 51–80.
Scannell, L., and R. Gifford. 2010. Defining Place Attachment: A Tripartite Organizing Framework. *Journal of Environmental Psychology* 30 (1): 1–10.
Schüßler, R. 1935. *Das Sternecker-Museum der NSDAP in München, Tal 54*. Munich: R. Schüßler.
Scobie, A. 1990. *Hitler's State Architecture: The Impact of Classical Antiquity*. University Park: Penn State University Press.
Sevilla-Buitrago, Á. 2015. Urbanism and Dictatorships: Perspectives of the Field of Urban Studies. In *Urbanism and Dictatorship: A European Perspective*, edited by H. Bodenschatz, P. Sassi, and M. W. Guerra, 27–35. Basel, Switz.: Birkhäuser.
Smith, J. S. 2018. Introduction: Putting Place Back in Place Attachment Research. In *Explorations in Place Attachment*, edited by J. S. Smith, 1–16. London: Routledge.
Spahn, B. 1937. *München: Hauptstadt der Bewegung*. Berlin: Riegler.
Speer, A. 1981. *Inside the Third Reich: Memoirs*. Translated by R. Winston and C. Winston. New York: Macmillan.
Spotts, F. 2003. *Hitler and the Power of Aesthetics*. Woodstock, N.Y.: Overlook Press.
Stephan, H. 1937. Zwei nationalsozialistische Feierstätten Segeberg in Holstein und Northheim am Harz. *Zentralblatt der Bauverwaltung* 57 (48): 1193–1197.
Strong, M. 2018. Influence of Memory on Post-resettlement Place Attachment. In *Explorations in Place Attachment*, edited by J. S. Smith, 19–32. London: Routledge.
Tamms, F. 1943. Die Kriegerehrenmäler von Wilhelm Kreis. *Die Kunst im Deutschen Reich* 7 (3): 50–57.
Troost, G., ed. 1938. *Das Bauen im Neuen Reich*. Bayreuth, Germany: Gauverlag Bayerische Ostmark [1943, 2d ed.].
Zum 9. 1942. November 1942: Gedenktag für die Gefallen der Bewegung. *Die neue Gemeinschaft* 8 (9): 492–502.

GEOGRAPHIES OF PLACE ATTACHMENT: A PLACE-BASED MODEL OF MATERIALITY, PERFORMANCE, AND NARRATION

ALEXANDER C. DIENER and JOSHUA HAGEN

ABSTRACT. Attachments are intrinsic to the human condition. Geographers and related scholars are especially interested in place attachments and how they shape everyday routines (e.g. routes to work, shopping, social interactions), major life choices (e.g. places of residence, education, vacations), and identities (e.g. civic, national, religious). This article explores the complexity and richness of place attachment as researched across multiple disciplines while also outlining how geographers can make a more pronounced contribution. To that end, this article surveys place attachment research and in the process demonstrates the recent dearth of explicit research on the topic in geography compared to other disciplines, especially psychology. Subsequent sections show how geographers implicitly engage place attachment as a theme, even if by other names, and identify areas of possible synergy with cognate disciplinary approaches. In an effort to reinvigorate interest among geographers and emphasize the need for a multi-dimensional consideration of place as integral to understanding place attachment, this article provides a model of place attachment as an assemblage of materiality, performance, and narration. Rather than being static or deterministic, this model points to people's varied capacities to make and remake place attachments, and in doing so reasserts the power of place in place attachment. *Keywords: place attachment, place identity, territorialization, environmental psychology.*

In the simplest sense, the concept of "place attachment" involves the study of people's emotional connection to place and how those connections shape thought and behavior. Today, research on place attachment is dominated by environmental and social psychologists employing a mechanistic epistemology disconnected from geographical research, which once greatly contributed to the field. Those researchers emphasize "attachment," commonly through interpersonal attachment theory, and tend to view "place" as a locational object of mental bonds determined by measurable and quantifiable attributes (i.e., age, social status, physical features, time spent, etc.). Geographers, when explicitly engaging the topic, generally focus on the dynamic and multi-dimensional aspects of place and the processes by which attachments form, disolve, and reform. As a result, geographers generally elide scalar fixity, immutability, and efforts to identify causal links between distinct inputs and outcomes (Adams 2014, 130-131). As a result of differences in terminology and approach, place attachment research in psychology, as well as cognate fields like sociology, anthropology, and political science, generally takes for granted the role of place and too often the perspectives of geographers.

This article contends that geography has much to contribute to place attachment research by approaching the topic through the discipline's varied epistemological, methodological, and theoretical frameworks. That variety could be considered a weakness, since those varied frameworks – sense of place (Tuan 1977), topophilia (Tuan 1974), place identity (Sopher 1979), place dependence (Meyer and Guss 2017), place meaning (Sack 1997; Antonsich 2010), rootedness (Tuan 1980), insideness (Relph 1976), homeland psychology (Kaiser 2002), sense of community (Agnew 1987; Kaplan 2018), place belonging (Gustafson 2006) – potentially dilute the concept and resist a synthesizing theoretical treatment. On the other hand, the varied perspectives within geography demonstrate the salience of place attachment research and its capacity to be broadly informed and applicable to understanding a range of issues (Manzo and Devine-Wright 2014, 2).

Rather than attempting to narrowly circumscribe place attachment, this article, like the articles of this special issue, highlights the potential of that breadth of perspectives in geography to contribute to place attachment research. The article begins with a review of place attachment in recent historical perspective that highlights the contributions of geographers. The subsequent sections demonstrate geography's current marginality in this burgeoning field and emphasize the array of topics pursued by geographers in which place attachment is implicit and might be more central. Finally, the article presents a place-based model that speaks directly to the multi-dimensionality of place attachment and points to future avenues of research wherein geographers can both learn from extant approaches and contribute in unique ways.

Origins of Contemporary Place Attachment Research

Marc Fried's 1963 study of Boston residents displaced through urban renewal is commonly regarded as social science's pioneering effort to study place attachment. Fried's research documented that residents experienced a sense of grief in being displaced from their neighborhood comparable to that accompanying the death of a loved one. In the 1970s and early 1980s, geographers took up the topic as central to the discipline's pivot away from spatial science. The work of Yi-Fu Tuan (1974, 1977, 1980), Edward Relph (1976), Anne Buttimer (1980), David Seamon (1980, 1984), and David Sopher (1972), among others, stimulated a reappraisal of place as integral to human experience. In doing so, they emphasized the multi-dimensionality of place that draws together social, symbolic, and natural elements (Tuan 1977; Seamon 1979).

Since the 1980s, the topic of place attachment was increasingly taken up by diverse disciplines bringing to bear specific scalar and epistemological foci. Environmental psychology's application of "attachment theory" (Bowlby 1969) and sociology's efforts to quantify and predict people-place bonds (Stedman 2002; Scannell and Gifford 2014) were pioneered by Irwin Altman (1975), Roberta Feldman (1990), Harold Proshansky and others (1970, 1983), and Ralph Taylor (1988), among others. These scholars accepted the general premise that differences in age or social status, for example, were sufficient to explain degrees of place attachment. Such assumptions rendered geographical concerns largely irrelevant. Altman and Low's 1992 edited volume *Place Attachment* was among the more textured treatments of the topic but contained not a single contribution by a geographer, for example.

For their part, human geographers in the late 1980s drew from recent social and critical theory to advance the "new cultural geography." These geographers tended to focus on culture's capacity to construct society in a processual sense, rather than simply reflecting it. By advancing a notion of culture as a medium through which personal and group identity is established, latent notions of place attachment's scalar fixity (i.e., home/homeland) and immutability likely rendered geography's explicit work on the subject less attractive. Yet while the term was rarely used in geographical research, "place attachment," broadly defined, was never absent from the discipline. The following sections demonstrate geography's limited, explict use of the term "place attachment" and emphasize its continual, implicit engagement with the concept.

Centers and Margins of Current Place Attachment Research

The most extensive literaure review on the topic of place attachment (Lewicka 2011) and more recent surveys of the field (Manzo and Devine-Wright 2014, 2; Giuliani 2016; Smith 2019, 5-6) generally acknowledge psychologists Leila Scannell and Robert Gifford's (2010) tripartite theoretical framework as the most popular perspective for conceptualizing place attachment. Scannell and Gifford divide place attachment research into studies generally centering on "person" (*who* is attached and why), "place" (*what* people are attached to in varied forms and scales), and "process" (*how* a place becomes meaningful and what impact this has on individuals), to paraphrase geographer Jeffrey Smith (2019, 5-6). Given the recent predominance of environmental psychology,[1] it is not surprising that research has focused most attention on the "person" aspect and measuring the intensity of place attachments (Lewicka 2011, 218; Hernandez and others 2014, 137). On this basis, geographers are uniquely positioned to advance understanding of both the "process" and "place" components of Scannell and Gifford's model by delving into the synchrony of nature, social relations, and meaning.

This would help combat reductionist tendencies in place attachment research, but since the late 1980s to the present, geographers have nevertheless drifted to the margins of the field (Smith 2019, 3). The term "place attachment" appears in the titles in 211 publications for Taylor & Francis journals and 72 articles for Sage, including several book reviews.[2] The compound search criteria of "place attachment" in the title and "geography" anywhere in the work resulted in 155 articles in Taylor & Francis journals and 34 in Sage, again including several book reviews. The Taylor & Francis subject-refining-filter enabled limiting the search to "geography journals'" (writ large) for articles with "place attachment" in the title, producing a list of 49 articles and book reviews.

Taking this a step further to search for "place attachment" in article titles in some key human geography journals, the results shows 0 in *Progress in Human Geography*, 1 in the *Annals of the American Association of Geographers*, 5 in *Geographical Review* (prior to this special issue), 2 in *Transactions*, 5 total in the *Environment and Planning* series, and 1 in *Geografiska Annaler*.[3] Not to belabor the point, but neither *Dictionary of Human Geography* (Gregory and others 2009),[4] *International Encylopedia of Human Geography* (Kitchin and Thrift 2009), *Oxford Dictionary of Geography* (Mayhew 2009), or *Oxford Bibliographies Geography* (Warf 2020) have entries for "place attachment." Only *Oxford Dictionary of Human Geography* (Rogers and others 2013) offers a specific entry for the term.

Although the term "place attachment" is rarely used in titles of geography articles, the topic has always been implicit in a broad spectrum of human geography research. Even when the term seemed the least prominent from the 1980s to the present, geographers worked consistently on the political, social, economic, and cultural conotations and significance of place. Their work commonly, albeit implicitly, spoke to the relevance of place attachment within broad ranging topics, such as borders, urban landscapes, diasporas, migration, aging, territoriality, sacred space, memory, indigeneity, education, health, gender, LGBTQ+, conservation, and citizenship. Place attachment, as the above survey of article titles attests, may not be at the fore of disciplinary terminology, but geography has consistently engaged it thematically and holds great potential to inform its study.

Synergies between Geography and the Place Attachment Research Community

Although far from comprensive, this section makes the case that geography and the place attachment research community overlap considerably across an array of topical and thematic interests.[5] One example is the field of GeoHumanities (see Cresswell and others 2015) and a variety works on poets' and writers' attachments to place as both thematically central or affectively influencing their writing (see Buell and others 2011; Burt 2015; Murphy 2018). Other examples of thematic intersection include "placelessness" approached most notably by geographer Edward Relph (1976), the effects of involuntary place separation and mobility (e.g., Fullilove 1996; Castells 1997), and "geographic belonging" considered by geographers like Daniel Trudeau (2006), Lise Nelson and Nancy Hiemstra (2008), and Marco Antonsich (2010), as well as other scholars like bell hooks (2009) and John Tomaney (2015). Kevin McHugh and Robert Mings (1996) and Alison Blunt and Robyn Dowling (2006) are examples of geographers engaging attachments to home, which resonate in place attachment research on aging, tourism, leisure activity, and second homes (Smith 2009; Gustefson 2014). Anthropolgical genetics (Crawford 2007) and genetic geographies (Nash 2015) both delve into the emerging complexities of people and place linkages encoded in DNA. Environmental cognition and behavior have a long history in geographical research

(Devine-Wright 2014) and have also been fruitfully engaged by other fields through reflection on place attachment (Stedman 2003; Carrus and others 2014). Geographic research on historic preservation, heritage, and identity is also well known (Hagen 2006) and pertinent to architects Dak Kopec and AnnaMarie Bliss's 2020 edited volume on place attachment, heritage, and authenticity. Terms such as patriotism, citizenship, nationalism, territorialization, homeland, and nation-state are thematic opportunities for political geographers to inform other disciplinary perspectives of place attachment (Kaiser 2002; Murphy 2013).

Equally so, social, urban, and economic geographers working on topics such as race, alienation, social class, bereavement, landscape, mobility, religion, place utility, etc. will find references to those themes in "place attachment" specific research: mobility in Gustafson (2006, 2014); social capital in Mihaylov and Perkins (2014); community size (Hamilton and others 2002; Manzo and Perkins 2006); socioeconomic and ethnic diversity (Billig 2006; Stolle and others 2008; Grief 2009), population density (Ardoin and others 2019), permeability of boundaries (Kirkness and Tije'-Dra 2017), urbanophillia versus urbanophobia (Fe'lonneau 2004), and the role of discursive practice (Di Masso and others 2014). This dynamic body of research suggests that "place attachment" is entering a fruitful period that draws from varied perspectives affording ample opportunities for geographers to contribute to unprecedented understanding of the topic (Smith 2019).

Many more synergies between geographic research and other disciplines exist in cognate terms like place meaning, sense of place, place identity, territorialization, etc. Moreover, as noted by geographer Paul Adams (2019), geography's diverse array of methodologies is well suited to the varied dimensions and complexity of place attachment. This is not to suggest that geography is the only discipline capable of informing the field; quite the contrary, the diversity of fields, themes, and topics to which place attachment is relevant suggests that the concept should not be circumscribed epistemologically or reduced to a singular ontological source. The next section offers an affective model for considering the varied elements that constitute place attachment.

A Place-based Model of Place Attachment

This model of place attachment portrays the integrative dynamism of place in which elements of nature, social relations, and meaning are assembled through both conscious and unconscious actions, experiences, and perceptions of individuals and groups. Specifically, this theory posits that *materiality/familiarity of place* derives from nature, *performance/partiality of place* derives from social relations, and *narration/memory of place* derives from meaning (Figure 1).

While readily associated with their respective realms, the material, perfomative, and narrative elements of place attachment converge at the relational junction of self/community and place. Rather than static, these assemblages are contingent, dynamic, and malleable, being affected by human agency, as well as

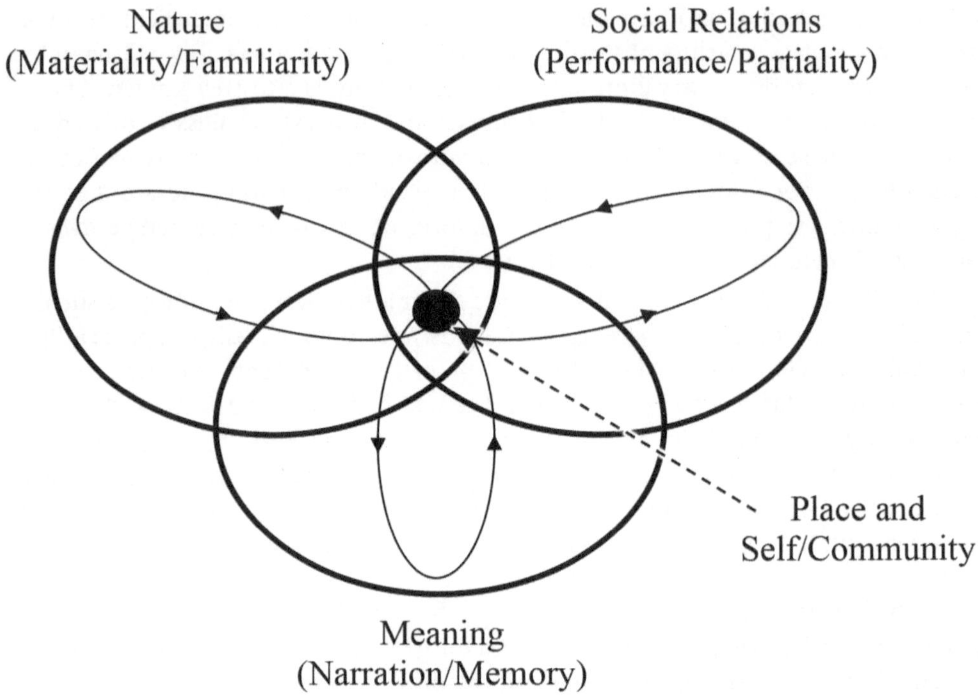

FIG. 1— Illustration of a place-based model of place attachment.
Source: Diagram created by Alexander C. Diener.

natural and social structures. As corporeal reality presumes a geographic setting, it is the synchrony of self/community and place that propels the assemblage and elucidates how activities, individuals, and communities come to be perceived or perceive themselves as site-specific, as attached to place.

MATERIALTITY: FAMILIARITY OF PLACE

Elements deriving, at least primarily, from the realm of nature constitute the materiality of place attachment; that is the stuff of the world or those things both animate and inanimate (including the human body) that are experienced in and through place. Like scholars across the social sciences and humanities, human geographers have focused more on social relations, as well as human cognition, perception, and intentionality, but a recent turn toward "new materialism" reasserts the corporeality and materiality of human experience and action. This is particularly pertinent to place attachment (Sack 1997; Malpas 2018).

According to Tuan (1977, 159), "attachment of a deep subconscious sort may come simply with familiarity and ease, with the assurance of nurture and security, with memory of sounds and smells, of communal activities and homely pleasures accumulated over time." Rather than a genetic or biological bond to place common in earlier theorizations of human territorialization (Ardrey 1966;